FOOD IN A CHANGING CLIMATE

D0770915

SocietyNow

SocietyNow: short, informed books, explaining why our world is the way it is, now.

The SocietyNow series provides readers with a definitive snapshot of the events, phenomena and issues that are defining our twenty-first century world. Written by leading experts in their fields, and publishing as each subject is being contemplated across the globe, titles in the series offer a thoughtful, concise and rapid response to the major political and economic events and social and cultural trends of our time.

SocietyNow makes the best of academic expertise accessible to a wider audience, to help readers untangle the complexities of each topic and make sense of our world the way it is, now.

Corbynism: A Critical Approach
Matt Bolton

The Smart City in a Digital World
Vincent Mosco

Kardashian Kulture: How Celebrities Changed Life in the 21st
Century
Ellis Cashmore

Reality Television: The TV Phenomenon that Changed the
World
Ruth A. Deller

Digital Detox: The Politics of Disconnecting
Trine Syvertsen

The Olympic Games: A Critical Approach
Helen Jefferson Lenskyj

Praise for *Food in a Changing Climate*

Food in a Changing Climate could not be more timely, as Covid-19 has revealed the enormous institutional vulnerabilities of the existing food system while the Black Lives Matter movement is propelling a long overdue reckoning with the insidiousness of racial capitalism. With impressive grounding in international scholarship, Alana Mann asks her readers to attend to the complex ecologies, cultures and political economies in which food is entwined and commit to a food politics that does not shy away from the difficult questions.

–*Julie Guthman, Professor of Social Sciences, University of California Santa Cruz*

Don't be fooled, this compact book speaks volumes to the civilizational crisis facing our societies – and to the strategies that can help us put our food systems back on track. *Food in a Changing Climate* brings together a wide range of data, information and expert opinion – as well as ancient wisdom – for a trenchant analysis of our dysfunctional capitalist food system. Can we feed the world with GMOs? Will fake meat cool the planet? Is the Blue Revolution the answer to overfishing? Alana Mann bravely takes on these issues in clear, no-nonsense language. Uncompromisingly honest, this book is a must-read for students of food studies and food activists seeking the facts and the language to speak truth to the power in our food system.

—Eric Holt-Giménez, Former Executive
Director of Institute for Food and
Development Policy/Food First

Wielding the food lens brilliantly, Alana Mann issues a wake-up call to the plunder of life-worlds and ecosystems at this geological tipping point. Her comprehensive account of planetary and species damage by industrial food, now intensifying claims to a future of lab-grown nutritionism, is exceptional. She brings her remarkable communication skills to critique the corporate scientism of food engineering and the urgency of restoring sovereignty to diverse food cultures in the illiberal shadow of standardisation. *Food in a Changing Climate* is a disturbing reminder of the plantation-like mindsets and practices of a globalized food system, and the need to replace it with an ethical world in which many worlds may fit sustainably.

—*Philip McMichael, Professor of Global
Development, Cornell University*

FOOD IN A CHANGING CLIMATE

BY

ALANA MANN
The University of Sydney, Australia

United Kingdom – North America – Japan – India
Malaysia – China

Emerald Publishing Limited
Howard House, Wagon Lane, Bingley BD16 1WA, UK

First edition 2021

Copyright © 2021 Alana Mann
Published under exclusive licence by Emerald Publishing Limited.

Reprints and permissions service
Contact: permissions@emeraldinsight.com

No part of this book may be reproduced, stored in a retrieval system, transmitted in any form or by any means electronic, mechanical, photocopying, recording or otherwise without either the prior written permission of the publisher or a licence permitting restricted copying issued in the UK by The Copyright Licensing Agency and in the USA by The Copyright Clearance Center. Any opinions expressed in the chapters are those of the authors. Whilst Emerald makes every effort to ensure the quality and accuracy of its content, Emerald makes no representation implied or otherwise, as to the chapters' suitability and application and disclaims any warranties, express or implied, to their use.

British Library Cataloguing in Publication Data
A catalogue record for this book is available from the British Library

ISBN: 978-1-83982-725-9 (Print)
ISBN: 978-1-83982-722-8 (Online)
ISBN: 978-1-83982-724-2 (Epub)

ISOQAR certified
Management System,
awarded to Emerald
for adherence to
Environmental
standard
ISO 14001:2004.

Certificate Number 1985
ISO 14001

INVESTOR IN PEOPLE

For the food and health workers

CONTENTS

ACRONYMS

ADM	Archer Daniels Midland
AFF	Alliance for Fair Food
AFM	Alternative Food Movement
AGRA	Alliance for a Green Revolution in Africa
AMR	antimicrobial resistant
ANAP	National Association of Small Farmers (Cuba)
AoA	Agreement on Agriculture
AOSIS	Alliance of Small Island States
ATSIA	Aboriginal and Torres Strait Islander Australians
CAFO	Concentrated Animal Feeding Operation
CCD	Colony Collapse Disorder
CDC	Centers for Disease Control and Prevention
CI	Conservation International
CIC	Consorzio Italiano Compostatori
CIW	Coalition of Immokalee Workers
CM	Cerrado Manifesto

CSA	Community Supported Agriculture
CSM	Civil Society Mechanism
DDT	Dichlorodiphenyltrichloroethane
ENSO	El Niño–Southern Oscillation
EOD	Earth Overshoot Day
FAO	Food and Agriculture Organisation
FDA	Food and Drug Administration
FIAN	Food First Information and Action Network
FSC	Federation of Southern Cooperatives (US)
FTA	Free Trade Agreement
GBR	Great Barrier Reef
GBRMPA	Great Barrier Reef Marine Park Authority
GDP	Gross Domestic Product
GFC	Global Financial Crisis
GHG	Greenhouse Gases
GMO	Genetically Modified Organism
GRSB	Global Roundtable on Sustainable Beef
HFCS	High Fructose Corn Syrup
ICTs	Information Communication Technologies
IFAD	International Fund for Agricultural Development
IMF	International Monetary Fund
IPCC	Intergovernmental Panel on Climate Change
ISI	Import Substitution Industrialisation
ISKN	Indigenous Seed Keepers Network

JIT	Just-In-Time
KFC	Kentucky Fried Chicken
LNG	liquefied natural gas
MACAC	Farmer to Farmer Agroecology Movement (Cuba)
MINT	Minimum-Input No-Till agriculture
MSC	Marine Stewardship Council
MST	Movimento dos Trabalhadores Rurais Sem Terra (Brazil)
NAFSN	New Alliance for Food Security and Nutrition
NCD	Noncommunicable disease
NFU	National Farmers Union (Canada)
OECD	Organisation of Economic Cooperation and Development
PCFS	The Peoples' Committee for Food Sovereignty
PETA	People for the Ethical Treatment of Animals
PICT	Pacific Island Countries and Territories
POPs	Persistent Organic Pollutants
PPE	personal protective equipment
PRAI	*Principles for Responsible Agricultural Investment that Respects Rights, Livelihoods and Resources*
RSPO	Roundtable on Sustainable Palm Oil
R&D	Research and Development
SAP	Structural Adjustment Program
SOFI	State of Food Security and Nutrition report
SoyM	Soybean Moratorium

TEK	Traditional Ecological Knowledge
TNC	Transnational Corporation
UNCTAD	United Nations Conference on Trade and Development
UNDP	United Nations Development Programme
USAID	United States Agency for International Development
USDA	United States Department of Agriculture
WCFS	World Committee for Food Security
WEIRD	Western, Educated, Industrial, Rich, Democratic
WFP	World Food Programme
WTO	World Trade Organisation
WWF	World Wildlife Fund

ACKNOWLEDGEMENT
OF COUNTRY

This book was written on the land of the Gadigal People of the Eora Nation, the coastal people of the area we now call Sydney. I acknowledge the traditional custodians of the lands on which I live and pay my respects to ancestors and elders, past and present. I honour Australian Aboriginal and Torres Strait Islander peoples' unique cultural and spiritual relationships to the land, waters and seas.

This land was never ceded.

ACKNOWLEDGEMENTS

In 2020 Earth Overshoot Day (EOD) landed on August 22, according to the Global Footprint Network. This is the day on which our consumption of natural resources met the Earth's ecosystem's capacity to renew across the entire calendar year.

It was a good year for the planet – carbon dioxide emissions from fossil fuel use were lowered. We logged timber at a lower rate, we traveled less. There was a 9.3% reduction in our Global Ecological Footprint compared to the same period in the previous year, 2019, when EOD fell on July 29.

But to come close to breaking even, we need to go back to consumption levels of 1970, when EOD fell on December 29.

This book joins a chorus of voices that have been demanding, for decades, that we take action to curb our impact on the environment. You include scholars, activists, and ordinary people. It is your work, your stories, and your ideas that make up this book. Thank you for sharing them. Any errors of interpretation are mine.

It is a rare privilege to be asked to write on such an important topic. I thank Jen McCall at Emerald for inviting me to contribute to a series that shares impactful, transformative research with a wide audience.

I am also indebted to University of Sydney graduates Drew Rooke and Justine Landis-Hanley who provided me with brilliant research support and access to their provocative thinking. To Drew, thank you also for your masterful editorial guidance, which kept me on track.

1

WE DIDN'T START THE FIRE

AN UNEASY STORY

Every year hundreds of *gestores de encomiendas* or 'parcel managers' transport millions of US dollars in cash and goods between El Salvador and the United States. This exchange, between those who fled El Salvador's 12 year civil war (1979–1992) and those who remain behind, fuels a 'reflective nostalgia' in which certain foods play a central role-*pupusas* (stuffed tortillas), *maíz blanco* (white corn), *frijoles de seda* (silk beans), *miel de caña* (sugarcane honey) and traditional unsalted *queso fresco*, a fresh farmers' cheese. One courier says 'what flourishes most, or what is above all else, is that feeling, those desires to want to eat and feel Salvadoran flavour, Salvadoran food' (cited in Anastario, 2019, p. 57).

These longings demonstrate how eating, one of our earliest and most instinctual behaviours, evokes memory and emotion, especially a sense of belonging.

> *Food is fellowship. Food is also nostalgia; our earliest meals can seed a sweet tooth, spoil an appetite, or*

*instil a craving later in life. The barrier between taste
and memory is paper thin.*

(Giggs, 2020, p. 215)

For the expatriate Salvadorans, familiar foods connect soil, crops, livestock and labour in an extension of local agricultural and social practices. Traditional healing teas and other plant-based remedies recall everyday care-taking behaviours and are essential to those without access to health care.

Food is central to the processes of dispossession, migration, transplantation and consumption that have literally trans-formed bodies, cultures and environments throughout the world, throughout history. Its countless stories are propelled by endless appetites. For coffee, tomatoes, chillies, pineapples, bananas, sugar and maize, only some of the foods that have created empires and triggered revolutions. The 'Columbian exchange' (Crosby, 2003) entirely transformed parts of Africa, Asia and Europe, driving industrialisation and expansion. King Sugar and other monocrops brought incredible wealth to a few, and misery to many, in a process Eduardo Galeano describes as the 'pumping of blood from one set of veins to another; the development of the development of some, the underdevelop-ment of others' (1971, p. 83). The reliance on singular staples created famines from Ireland to India, leading to a Green Rev-olution based on genetically modified seeds, agri-chemicals and fossil fuels which are now directly responsible for climate change. In response, the colonial project continues in a land and resource grab justified by a sustainable development paradigm that now includes 'green' biofuels in its agenda for expansion. Under the racialised double-standard of development, the appropriation of resources and embodied labour from the Global South to high-income nations continues; in 2015, a net total of over 10 billion tonnes of materials and 370 billion hours of human labour (Hickel, 2020). This is the blood that continues to flow from the open veins of the Global South where 'resource rebels' (Martínez-Alier, 2003) who rely on and steward the environment for their livelihoods not only fight for their own survival but also that of the planet against 'a fossil fuelled imperialist drive for control and power over resources' (Wedge cited in Hayman, 2018, p. 82).

It is naïve to deny awareness that a food system built on extraction and exploitation was bound to fail; after all history shows us 'colonists might come to conquer but in the end they struggle with the inevitable impact their environment and situation will have on them' (Behrendt, 2016, p. 193). The evolution of global foodways is not an easy story. We know that 'stories are powerful tools and can be even more powerful weapons in the hands of malignant narcissists' (Yunakporta, 2019, p. 129). Told straight, the story of food in a changing climate challenges the 'inequalities, alienation, and violence inscribed in modernity's strategic relations of power and production' (Moore, 2015, p. 170) – the same forces naturalised in the popular Anthropocene narrative.

BREADBASKETS AND BASKETCASES

As I write, the world is reeling. The no longer so novel pathogen COVID-19 defies containment across cities and continents. Global food supply chain vulnerabilities are foregrounded as farmers and processors grapple to understand 'who needs what, who has what' to avoid farm closures and prevent food waste. Fresh milk is flowing down drains, thousands of animals are being culled and fruits and vegetables are rotting in fields and shipping containers (Cagle, 2020). Our newly crowned 'essential' food and farmworkers, many already engaged in campaigns for decent wages, fair working conditions and basic safety, are now challenged to socially distance with a shortage of personal protective equipment (PPE) if they can work at all (Arantini, 2020; IPES-Food, 2020). People are panic-buying in even the most affluent cities around the world, where rising unemployment is driving a demand for food banks which have never remedied endemic food insecurity (Power, Black, & Brady, 2020). Even more dire is the impact of the pandemic on nations already experiencing acute food insecurity because of a lethal combination of conflict, macroeconomic crisis, weather-related shocks and pest invasions. These include Afghanistan, Pakistan, Northern Nigeria and the Democratic Republic of the Congo (DRC), and much of East Africa where the worst locust invasion in 70 years is decimating crops and livestock feed (Smith & Kayama, 2020). Yemen, the most food insecure nation in the world with 53% or 15.9 million citizens in crisis, is at Catastrophe (Phase 5) level after three years of civil war. We lack data for many other countries, such as Iran and the Philippines (Ghosh, 2020), where the virus is used as an opportunity for regimes to crackdown on personal freedoms.

Rising food prices can be directly attributed to the impacts of movement restrictions and illness on local markets and the unavailability of agricultural labour, but the roots of the problem extend much deeper. The pandemic has exposed the fragility of a food system built to rely on interconnected and complex global supply chains that facilitate trade between nations. Based on the logics of comparative advantage, the 'free' trade regime dictates that countries can import all the food they need if that is cheaper than growing it at home. For example, if Egypt can't grow wheat as cheaply as they do around the Black Sea, it should import a large amount of wheat from Russia to supplement its own small harvest in feeding its population. That works well when harvests are high and conditions for transport are optimal but not so well when extended droughts in Russia lead to higher prices, financial speculation on harvests and stockpiling of grain by competing nations. Russia has a history of imposing restrictions and taxes on wheat, and following destructive droughts in 2010 completely banned trade, contributing to the Arab Spring uprisings. Climate change was a decisive spanner in the works of the global food trade well before the pandemic hit.

A glaringly obvious impact of food dependency, highlighted by the pandemic, is that when supply chains fail, people starve. If local farmers stop growing, their families don't eat, and without jobs they cannot buy food. Put simply, for all the advantages of specialisation in growing, processing and distributing specific foods, the efficiencies gained come at the expense of social goals including food security, the need to preserve livelihoods and the protection of the environment (Clapp, 2020). Our obsession with efficiency has decimated local and diverse food economies in many parts of the world by closing down small farms, regional abattoirs and farmers markets, eroding local resilience. Now, without adequate inventories and regional self-sufficiency, our cities are 'nine meals from anarchy' when crisis strikes (Fraser, 2020).

To maximise profits and keep costs low, food corporations have applied the principles of 'just enough, just in time', balancing supply and demand. Disruptions like closed borders, trade and visa restrictions, and limited distribution channels expose the precarious nature of this just-in-time (JIT) model. In policy terms, our food system is 'locked in' to the point where 'social forces and decisions can reinforce a lack of change or compound failure'. As such, rather than sustaining us as we grapple with COVID-19 'how the food system operates is a significant threat to the ecosystem's future and humanity's within that' (Lang, 2020, p. 197).

This book was always going to be about the need to adapt our food behaviours in response to societal collapse driven by climate change. The coronavirus has added a new urgency to this narrative. Climate change interacts with and exacerbates existing inequities including those in our health systems. Through the pandemic lens, we witness how the most vulnerable are those most affected by the virus and its profound economic impacts – the elderly, people of colour, the jobless, the refugees of failed states on which we have little or no data regarding infections and deaths. In 2014, the World Health Organisation (WHO) warned that the most significant health risks and impacts of climate change would not be experienced equally among nations, regions or groups of people within metropolitan areas (WHO, 2014). The cracks are showing as countries with poorly resourced health systems, and those with the highest rates of inequality, are crippled by the pandemic. Societal collapse has arrived. And with it a renewed focus on not just our food systems but the way we live.

FACING THE DRAGON

I started writing this book in the midst of a more local, but as personally devastating, national tragedy – the worst bushfire season in living memory. As a child in mid-70s rural Australia, my class was set the book *February Dragon*. Author Colin Thiele perfectly captured the terror of a community faced with an uncontrollable bushfire. In 2019, the dragon came early, invited by climatic changes including extended drought and the positive Indian Ocean Dipole which is contributing to acute food insecurity across East Africa (Marsham, 2020). Between October and January, more than 150 bushfires burned across Australia. Firefighters were unable to contain 64 of these, in some cases for months. Only heavy downpours were able to tame the flames. More than 18 million hectares burned, including 1.3 million of agricultural land and 1% of all vineyards. Over one billion animals and 34 people were killed, and thousands of properties razed. A further 445 deaths were attributable to smoke from the bushfires (Wahlquist, 2020). While the flames may have ceased the grief has not; as a nation, we are still in mourning.

The Australian bushfire season of 2019/2020 is a lesson in political paralysis and policy failure for a rapidly warming world. It incinerated much of but, astonishingly, not all of the doubt, apathy and indifference about climate change retained by policymakers and the powerful elite in Australia and around the world. Coupled with the devastating impacts of the pandemic, it feels like collapse has begun, and the time to embrace what Jem Bendell (2018) calls 'deep adaptation' has arrived.

In 2018, when this concept went viral with his paper *Deep adaptation: A map for navigating climate tragedy*, it exposed the ultimate inconvenient truth – societal collapse, defined as the end of familiar modes of sustenance, security, pleasure, identity, meaning and hope, is inevitable. Arguably, Bendell's view is supported by the Intergovernmental Panel on Climate Change (IPCC) declaration of a climate emergency in which we have perhaps 10 years to stall a tipping point or 'precipice' (Ord, 2020) that leads to catastrophe, and possibly human extinction.

According to the latest climate science, global warming is more rapid and severe than previously anticipated. Particularly at threat is worldwide food security, and with it well-being and social stability. Climate change is predicted to lead to a decrease in food production. Tropical zones will move from optimal growing conditions for cereal crops like rice and corn into extreme and prolonged summer temperatures. This will cause drops in productivity in areas where the bulk of malnourished people already live. Growing seasons will likely get longer in temperate zones as climate warms but any gains will be offset by extreme weather events like Australia's bushfires, which followed the longest drought in living memory.

Climate-related droughts across the globe this decade alone have caused drops in wheat production of 33% in Russia, 19% in the Ukraine, 14% in Canada and 9% in Australia (Vidal, 2013). The disruption of immense ocean currents is already impacting on fisheries which are increasingly exploited for the protein needs and shifting tastes of a global population expected to reach nearly 10 billion by 2050. To feed this population sustainably, the World Resources Institute claims we need to close three gaps: a 56% food gap in 'crop calories'; a 593 million hectare 'land gap' (an area twice the size of India); and an 11 gigaton greenhouse gas (GHG) mitigation gap between anticipated agricultural emissions in 2050 and the target level we need to hold global warming below two degrees Celsius and avoid catastrophic climate impacts (Ranganathan, Waite, Searchinger, & Hanson, 2018). With half the Earth's surface already under production for food and even more for fuel crops, we are running out of land – a problem exacerbated by rising seas. A one-foot rise in sea level is sufficient to consume Tuvalu, Kiribati and the Maldives in the Pacific, Kivalina in the Arctic and coastlines around the world from Miami to Mumbai. By 2100, seas are expected to rise three feet (Mann & Toles, 2016). Storm surge is a major risk to health and food supply in urban areas, particularly where critical infrastructure is situated on the coast. Hurricane Sandy proved the vulnerability of low-lying coastal areas in New York, pouring 1.6 billion gallons of untreated sewage into local waterways. The peninsula Hunt's Point, the poorest congressional district in the United States and the source of half of the city's meat and fish and 60% of its produce, was only spared in Sandy through a 'happy accident of the tides'; next time the city may not be so lucky (Glickman, 2020).

Climate inequality will generate waves of migrants and refugees on a scale we have never seen. While migration and climate have always been connected, 'the impacts of man-made crisis are likely to extensively change the patterns of human settlement' (UN, 2019). In 2018, 17.2 million people were displaced through climate-related disasters (IDMC, 2019). Before the pandemic, the World Bank projected internal climate migration in the order of 143 million people by 2050 in just three regions: Sub-Saharan Africa, South Asia and Latin America (Rigaud et al., 2018). These areas, the most vulnerable to slow-onset climate impacts like water stress, crop failure and sea level rise, will host local 'hotspots' of climate in- and out-migration as people move from coastal and rural places to urban and peri-urban centres, placing increasing pressure on housing and transport infrastructure, social services and employment opportunities.

Compounding these challenges, the way we produce food is a significant contributor to global warming. The IPCC reports agricultural production is complicit in increasing emissions, reducing biodiversity and polluting environments. The entire food production system, including transportation and pack-aging, is responsible for as much as 37% of total GHG emissions (IPCC, 2019). At the food-energy-water nexus, agriculture is already in competition for water with extractive activities like coal mining and the hydraulic fracturing or 'fracking' of non-traditional fossil fuels like shale and coal seam gas as we exhaust more accessible hydrocarbon resources (Wright & Nyberg, 2015). One third of the food we produce globally is wasted – a 'double waste' of energy in terms of non-consumed food energy and the inputs required for production, transport and distribution (Vittuari, De Menna, & Pagani,

2016). This 'hidden burden' is compounded with COVID-19 as restrictions on movement, road and port closures, and limited access to markets reduce the quantity and quality of food, especially perishables (FAO, 2020a; FIAN International, 2020a).

The IPCC applies sustainability criteria in its analysis of the impacts of food production and consumption on the planet. It recommends that

> ...balanced diets featuring plant-based foods such as those based on coarse grains, legumes, fruits and vegetables, nuts and seeds and animal-sourced food produced in resilient, sustainable and low-GHG emission systems present major opportunities for adaptation and mitigation while generating significant co-benefits in terms of human health.
>
> (IPCC, 2019, p. 27)

These particulars are a luxury for citizens of 'food bank nations' reliant on welfare or low wages that are insufficient to put food on the table, also for Indigenous communities dispossessed of their traditional lands and foodways (Riches, 2018). Emissions reduction is also secondary to survival for the 800 million chronically hungry who do little to contribute to global warming in the first place. In the same year as the release of Bendell's paper, more than 113 million people across 53 countries experienced 'acute hunger requiring urgent food, nutrition and livelihoods assistance' (FSIN, 2019). The World Food Programme (WFP) warns that another 265 million may be driven into acute food insecurity by COVID-19 (Beaumont, 2020).

This is more than another food crisis. The language of crisis offers us a 'deceptive optimism' by leading us to feel that 'we are simply faced with a perilous turning-point with an imminent outcome, or even an opportunity' (Bonneuil & Fressoz, 2017, p. 21).

More significantly, it shares with discourse about anthropogenic climate change a focus on 'the *novelty* of crisis rather than being attentive to the historical *continuity* of dispossession and disaster caused by empire' (DeLoughrey, 2019, p. 2). Without this backstory, any analysis of food in a changing climate 'misses the globe' and those at the forefront of climate change.

While we quibble over language, stored carbon in the atmosphere will continue to contribute to global warming even if emissions are cut to zero tomorrow. And they will not be; while the global economic shutdown forced by COVID-19 is anticipated to generate a drop of 6–8% in emissions (IEA, 2020), about 7.6% *per year* is required to keep global warming under 1.5 degrees Celsius. If we learn anything from these data, it is that 'fiddling around the edges' of carbon transition through changing our personal travel and consumption habits and even moving to a no-growth economy is aimless without 'huge, lasting changes in energy systems' – especially our reliance on fossil fuels (Richard Betts cited in Vince, 2020, p. 9).

Here I am assuming that my readers believe that global warming exists. If not, I recommend as a primer Michael Mann's *The Madhouse Effect: How Climate Change Denial is Threatening Our Planet, Destroying Our Politics and Driving Us Crazy*, illustrated by Tom Toles (Mann & Toles, 2016). Mann lays out the science clearly and then addresses the war being waged against it by politicians, corporate elites and specific media outlets. He asks 'when is the right time to talk about climate change?'

A second reading, one that captures the cognitive dissonance of climate change denial with dark humour, is William T. Vollmann's 'chronicle of self-harm' *Carbon Ideologies*, described by *The Atlantic* as 'the most honest book about climate change yet' (Rich, 2018). Including nuclear power, solar energy, coal mining, oil extraction, fracking, tidal electricity and wind turbines in his wide critique, Vollmann's message to a reader on a 'hotter, more dangerous and biologically diminished planet' is 'nothing can be done to save [the world]; therefore, nothing need be done' (Vollmann, 2018, p. 3). His project is to reveal our human nature; our ability to 'not only sustain, but *accelerate* the rise of atmospheric carbon levels, all the while expressing confusion, powerlessness and resentment'.

Deep adaptation is all about mindset. Before we can undergo processes of deep adaptation in our foodways, we must locate ourselves, and our thinking, in relation to the current threat. Timothy Morton (2013) describes global warming as a hyperobject. These are objects 'massively distributed in time and space relative to humans', and the ghosts of actions past. In the menacing shadow of hyperobjects, contemporary decisions to ground ethics and politics in forms of process thinking and relationism are profoundly challenging. Morton says our dominant, market-based economic system is not capable of withstanding hyperobjects as it is reactive (consumer demands must be met!) rather than proactive, leaving us ill-equipped to deal with ecological fragility and societal collapse.

Food systems, invisible and omnipresent, share many characteristics of the hyperobject. Their transformation, as a vital element of action on climate change, demands

> ...theories of ethics that are based on scales and scopes that hugely transcend normative self-interest theories, even when we modify self-interest by many orders of magnitude to include several generations down the line or all existing lifeforms on Earth.
>
> (Morton, 2013, p. 138)

This revelation is impoverished in comparison with Indigenous conceptions of time. The Anishinaabe people of the Great Lakes Region in North America are taught to plan seven generations into the future. Anishinaabe Ojibwe man Martin Reinhardt describes how

> ...*the circles of subsequent generations spiralling through time offer some assurance that our way of life will endure if we take responsibility for our actions today and pass along our valuable teachings.*
>
> (Reinhardt, 2015, p. 84)

Based on this philosophy, ancestors alive before European colonisation have a deep concern and investment in how their people eat and live today, and will into the future.

To take on this level of responsibility requires adopting what Apalech poet, artist and author Tyson Yunkaporta calls 'cultural humility' (2019, p. 98) in respect to others, including the many species we share the planet with. It also calls for us to reject hubris and defy reductionist thinking that suggests we can fix the planet by changing our diets or that we are insured by our embeddedness in the world food economy, an asymmetrical, volatile beast that inflicts a 'slow violence' on those it exploits. This is 'a violence that occurs gradually and out of sight, a violence of delayed destruction that is dispersed across time and space' (Nixon, 2013, p. 2). Its immense power is its *attritional* quality. Manifesting in heatwaves, famines and rising oceans, slow violence is 'incremental and accretive, its calamitous repercussions playing out across a range of temporal scales'. This rarely comes to the attention of the typical WEIRD (Western, educated, industrialised, rich, democratic) consumer, captured by the endless race for stuff and distracted by more vivid, spectacular episodes of violence that erupt in other places before quickly fading to black. Until the dragon arrives, and sets your country alight.

'We' didn't start this fire. But does that matter when no one can put it out?

THINKING ABOUT AND LIVING IN (FOOD) SYSTEMS

Climate change will compound our existing challenges as communities. It will exacerbate long standing inequalities in access to healthy food, clean water, fair incomes and safe environments, all of which are key elements of food systems. Simply put, a food system is a system in which food is grown, produced, processed, transported, consumed and wasted. More accurately it is 'an entire array of ideas, institutions, and policies that affect how food is produced, distributed, and consumed' including the roles of nutrition science, environmental regulation, agricultural research, international food aid and trade (Guthman, 2011, p. 19). This whole-systems perspective of our foodways reveals the entanglement of public health, politics and ecosystem destruction.

There are countless versions of food systems across the globe, and there have been many more throughout human history but the most dominant one, in terms of its contribution to global emissions, general misery and corporate profits, is the industrial or corporate food regime (McMichael, 2009). This behemoth crosses international borders and is dominated by multinational corporations supported by local and global governing frameworks. The 'Big Six' – BASF, Bayer, Dow Chemical, DuPont, Monsanto and Syngenta – have long dominated the sale of seeds and agricultural chemicals including pesticides and seed treatments. Recent megamergers between Bayer/Monsanto, Dow/DuPont have raised concerns with the US Department of Agriculture (USDA) over the implications of market concentration and reduced competition, especially for struggling farmers (Mann, 2019; McDonald, 2019). Most toxic are development assistance arrangements that criminalise traditional seed saving practices in countries like Tanzania, for example. In this case, legislation was modified to give commercial investors access to agricultural land and intellectual property rights as a condition for receiving foreign aid through the New Alliance for Food Security and Nutrition (NAFSN), launched by the G8 in 2012 with the assistance of the World Bank and the Bill & Melinda Gates Foundation (Daems, 2016). The Gates Foundation's Alliance for a Green Revolution in Africa (AGRA) is already widely criticised for 'failing on its own terms' to double productivity and incomes in 30 million small-scale farming households across the continent. In the 13 countries it focuses on the number of undernourished has increased 30% since the inception of the program (Wise, 2020). This is in part because of the replacement of more nutritious and climate resilient traditional crops like millet and sourghum with AGRA-sponsored priority crops such as maize.

In food production global giants JBS of Brazil, Tyson Foods Inc. in the United States, and WH Group of China dominate meat and dairy, controlling 63% of pork packing, 46 of beef, and 38 of poultry in an oligopoly, defined as a market where four firms control 40% or more of sales (Howard, 2019, p. 31). Tyson Foods Inc., Nestlé and Cargill produce emissions to rival those of fossil fuel companies and the top 20 global meat producers collectively emit more GHGs than Germany. Yet most lack mitigation plans and only four of the top 35 report on their emissions (GRAIN, 2019). Horizontally and vertically integrated, each firm controls dozens of brands, giving the consumer the impression of diversity. They expand concentrically by processing additional species (for example, Cargill is now moving into aquaculture with the purchase of Norwegian feed manufacturer EWOS), horizontally by acquiring competitors, and vertically by taking over upstream suppliers and downstream retailers. Hidden from view they are morphing a supply chain once constructed in stages and markets, each composed of diverse competitive firms, into a seamless system where a handful of firms control the entire supply chain (Howard, 2019).

These processes don't merely shape the way we produce, process and consume food; they influence global geo-politics. Agricultural systems in the Global South were transformed to produce cash crops like coffee and cotton for export, in doing so reducing access to land and other productive resources, including forests and fisheries, and wages (McMichael, 2010). Agribusinesses emerged to provide inputs for monocultural crops in the form of petroleum-based fertilisers, pesticides and farm machinery. This system laid the foundations for the paradox we face today – excessive production through fossil fuel-driven agriculture coupled with the vulnerability of the majority world to food insecurity. Our carbon economy, based on short-term goals and the poorly regulated exploitation of the planet's resources, leaves us utterly unprepared for survival in a warming climate. In what Amitav Ghosh (2016, p. 111) calls 'the Great Derangement', our lives and our choices are 'enframed in a pattern of history that seems to leave us nowhere to turn but toward our self-annihilation'.

The global food system is the poster child of an unregulated advanced consumer capitalism that is remaking 'both ecologies of the planet and the body …in ways we don't entirely know or understand' (Guthman, 2011, p. 195). To break this lock-in we need to understand capitalism. Eric Holt-Giménez (2017, p. 14) notes critiques of capitalism were anathema to elites before the global financial crisis (GFC) as 'even a perfunctory examination of capitalism immediately uncovers profound economic and political disparities, thus contradicting the commonly held notion that we live in a classless, democratic society'. Given the state of the world, the logic of neoliberal capitalism, characterised by the privatisation of public goods and the loosening of regulatory constraints on markets and corporations, is a fiction. To have faith in the market is to 'continue to adhere to the fable of the freedom given to each to choose his or her life' (Stengers, 2015, p. 29).

It has been said that 'capitalism can no more be "persuaded" to limit growth than a human being can be "persuaded" to stop breathing' (Bookchin, 1990). Yet this climate is also changing. The world is at a 'critical turning point – economically, politically and environmentally' (McMichael, 2020, p. 28) and worship of Great God Growth is now widely challenged. Disappointed too often, peoples around the world are exhibiting

> ...a growing reluctance to turn with confidence to the scientists and technologists, not to expect from them the solution to problems that concern the development they have been so proud to be the motor of.

> (Stengers, 2015, p. 29)

As we finally embrace the portents of climate science and face the prospect of collapse, are we turning to a different ethical-moral-governance paradigm to save not just ourselves but our planet?

WINNERS AND LOSERS

Homo sapiens is described by Yuval Noah Harari as the 'deadliest species in the history of planet earth...an ecological serial killer' (Harari, 2019). The announcement of a new geological era, the Anthropocene, locks in this hubris. It continues the mythology that 'man is the only animal to make tools; that man is the only animal with language, a sense of fairness, generosity, laughter; that man is the only mindful creature' (Rose, 2019, p. 55). It heralds the arrival of humanity as a geological force, 'both heroic and unsustainable, arousing both admiration and terror which reinforces a certain number of socio-environmental injustices under the consensual banner of the species' (Bonneuil & Fressoz, 2017, p. 93). Of course this consensus does not exist. At a time when the United States and the United Kingdom are shrugging off the 'burdens of solidarity' (Latour, 2019), 1% of the richest people on Earth possess more than the rest of the planet. Over the past 30 years, the growth of incomes of the bottom 50% has been zero while the incomes of the top 1% have grown by 300%. Eight men own more than the poorest half of the world's population (Oxfam, 2017).

While not by design, this is not by accident. The roots of this disparity germinated in 'colonial capitalism', which Macarena Gómez-Barris describes as

> ...the main catastrophic event that has gobbled up the planet's resources, discursively constructing racialized bodies within geographies of difference, systematically destroying through dispossession, enslavement, and then producing the planet as a corporate bioterritory.
>
> (Gómez-Barris, 2017, p. 4)

The colonial conceptualisation of territory as commodity continues to drive the modern food system. Its origins date back to 1492 and the genocide of Indigenous peoples and the destruction of their agricultural systems in the Americas; it is rooted in the violent transplantation of African and Caribbean peoples in the transatlantic slave trade (DeLoughrey, 2019). Today it is mirrored in the continued exploitation of an ever-shifting periphery that includes 'hotspots' for primary crops and livestock for international consumption (Sun, Scherer, Tucker, & Behrens, 2020) and sites for new extractivism, including the sea bed. In the most developed economies on Earth it persists in systems of 'racialized social control' in the form of 'laws, policies, customs and institutions that operate collectively to ensure the subordinate status' of groups of people 'defined largely by race' (Alexander, 2012, p. 13).

Given this fraught backstory, the announcement of the arrival of the Anthropocene as 'a moment of rupture of the temporality of modernity' misrepresents the modern era as 'monolithic and total' (Swyngedouw & Ernstson, 2018, p. 9). It creates an impression of before and after, temporally and geologically, that masks the reality of modernisation as an 'internally fractured' and highly contentious political process replete with inequalities. Not all of us gained by undermining Earth's boundary parameters. There is no universal 'we' behind the grand façade of the Anthropocene. 'We have never been Anthropos' claim those of us arguing from Indigenous, feminist and post-colonial perspectives (DeLoughrey, 2019). The losers are out of sight, off-stage. Among them are the 'stuffed and starved' (Patel, 2009) and the animal victims of the food system including birds, mammals and fish raised in feedlots. Anthropocentrism's myopic narrative lens conceals these unsightly realities (Mueller, 2017, p. 45).

The 'Anthropocene body' (Bonneuil & Fressoz, 2017) is the product of a consumerist society that has been moulded to accept, and crave, a dietary model high in meat, sugar and ultra-processed foods. The WFP reports 212 million chronically food-insecure and 95 million acutely food-insecure people (WFP, 2020), while the number of overweight and obese people rose from 857 million in 1980 to 2.1 billion in 2013 (Ng et al., 2014). Ultra-processed foods, classified as 'NOVA 4' by Brazilian public health nutritionist Professor Carlos Monteiro, are 'industrial creations, no different from paint or shampoo... designed to appeal to the consumer's palate' (Raubenheimer & Simpson, 2020, p. 135). Their raw materials include industrially farmed, high-yield crops, many of which are monocultures yielding high GHG emissions. Other ingredients are derived from the petroleum industry. This makes sense as Big Food and Big Oil share not only 'challenges and interests...but also the materials and processes that are used to solve the manufacturing challenges' (p. 136).

The by-products of late-capitalism, including chemical and plastic waste, fuel an economic regime of disposability captured by the Great Pacific Garbage Patch, a 'plastisphere' larger than the State of Texas (DeLoughrey, 2019, p. 101). These toxic substances enter our waterways and food chains on the basis of cost/benefit analysis as regulators determine 'reasonable' or 'acceptable' doses for foodstuffs and maximum concentrations in a devilish accounting engineered for economic gains. Even our towns, cities and suburbs are designed to serve the consumerist society where we drive to non-spaces like shopping malls – increasingly desirable destinations in a warming climate where escaping from the heat will become a daily mission. Meanwhile, neoliberal ideologies of individual responsibility and consumer freedom of choice deflect attention from the failure of elites to address the profoundly damaging and discriminatory externalities of production (Guthman, 2011; Scrinis, 2013).

The other bodies who lose in the global food system are the multispecies exploited to satisfy our growing appetites for all foods, ranging from beef to almonds. The easy targets, and justifiably so, are the industrial slaughterhouses and Concentrated Animal Feeding Operations (CAFOs) that make up 15% of 450,000 factory farms in the United States. In these plants, workers process 6,000 cattle or 20,000 pigs per day, or 140 chicken per minute in a state of 'hypnotic numbness that can be produced by concentration on the blur of constant motion from the constantly moving line of carcasses' (Pachirat, 2013, p. 217).

Realising that the well-publicised violence of industrial killing is alienating WEIRD consumers, the giants of the meat industry are now reinventing themselves as 'protein producers' by moving into plant-based and cellular meat production. This growing sector is adulterated by ultra-processed products that demand endless fields of monocultures like soy and almonds at the expense of soil health and biodiversity. While no and low meat diets reduce pressure on land and, ideally, generate crops with nitrogen fixing potential, they rely more on pollinator abundance and diversity, and increase impacts on freshwater ecosystems (Laroche, Schulp, Kastner, & Verburg, 2020).

Even discounting the massive environmental impact of factory farms and monocultures on multispecies and ecosystems, basic common sense alone dictates that they can not be part of a healthy, equitable food system. Not supporting these industries through our food choices is a no-brainer. If we can afford to eat outside the cruel and corrupt global supply chains that generate billions of dollars for Big Food, we should. It is short-sighted, however, to imagine that their practices will be brought to a halt by political consumerism or 'small-p' politics (Kennedy, Johnston & Parkins, 2018). We cannot shop ourselves out of the lock-in.

THE HIDDEN COSTS OF CHEAP FOOD

Hunger and obesity co-exist as a consequence of an industrial food system focused on overproduction of cheap food, where the costs are externalised in illness and environmental harm. A constant supply of cheap food is the 'bedrock of modern society' (Newman, 2019, p. 134). Legitimised by the myth of scarcity (Mann, 2017), it has driven expansionism in territory and profits from the days of overharvesting wild foods like the passenger pigeon – now extinct, but once 'free or for pennies' – to the celebration of the monstrous three-bird roast or 'turducken' (Newman, 2019, pp. 136–138). Technology-driven productiv-ity gains are framed as solutions to global food insecurity without corresponding reflection on their core tensions with contemporary consumption trends that are grossly inefficient and promote diets that are 'nutritionally derelict' (Taylor, 2015, p. 105). Small-scale farmers and consumer health are sacrificed to support a 'modernization narrative that appears unable to question the consumption patterns it serves'.

In WEIRD societies, the average eater spends just 10% of her income on food and, for decades, Big Food has had us believe we 'have never been safer, better fed, or healthier' (cited in Souder, 2012, p. 357). We receive these reassuring messages at the supermarket which offers us the illusion of safe, nutritious food. The aisles where the modern human forages are full of products masked by a 'nutronal façade' constructed by food manufacturers for marketing purposes (Scrinis, 2013), aided by 'choice architecture' that influences shopping behaviour by making products more desirable or salient through shelf position, colour, information and ordering of products (Thorndike & Sunstein, 2017). The best-sellers, located at point-of-sale in conventional supermarkets, include sugar-sweetened beverages, potato crisps and confec-tionery. Many of these products contain slow-acting toxins as deadly as many of the poisons found in the natural environ-ment which animals and traditional societies learnt to avoid. Ultra-processed foods containing these toxins, masked with flavours, give us positive flavour-feedback signals known as the Dorito Effect (Schatzker, 2015) and lead to the on-set of chronic illnesses like diabetes.

The emergence of diabetes reveals how the individualisation and de-politicisation of non-communicable disease supports the commodification of food for profit. A metabolic imbalance of insulin and high blood sugar, diabetes was one of the first known diseases to be documented. Symptoms were reported in Egyptian manuscripts in 1500 BC. This case was likely Type 1, an inherited autoimmune disease. Type 2 is a consequence of Western diets and lifestyles. Before COVID-19, diabetes and obesity were our most serious public health issues. In 2014, the McKinsey Global Institute stated that obesity alone is responsible for around 5% of all global deaths, and costs the international economy US$2 trillion or 2.8% of GDP annually, equivalent to the combined impact of armed violence, war and terrorism (Dobbs et al., 2014, p. 1). In 2017, the costs of diabetes treatment in the USA amounted to US$327 billion – four times the total profits of farmers that year (Metelerkamp, 2020). Diabetes disproportionately affects Indigenous populations including Aboriginal and Torres Strait Islander Australians (ATSIA) who are 1.2 times more likely to be overweight and 1.6 times more likely to be obese than non-Indigenous populations. While few Australians meet the dietary recommendations for healthy foods, ATSIA obtain 41% of their energy intake from discretionary foods high in saturated fat, sugar and salt, and consume 25% more sugar per day than other Australians (Gwynn, Searle, Senior, Lee & Brimblecombe, 2019).

Big Food notes that public concern about obesity, along with new taxes and additional government regulations, may reduce demand for their products. In response they sponsor campaigns, organisations and research that allege to fight the obesity epidemic. From 2011 to 2015, Coca-Cola Company and PepsiCo sponsored 95 national health organisations in the United States yet lobbied against 29 public health bills aiming to reduce soda consumption (Siegel, 2016). Rather than challenge Big Food, governments retreat to the language of personal responsibility, blaming so-called 'lifestyle related' diseases like diabetes on 'noncompliant' individuals. This is a 'broad exoneration' of the structures and systems that are implicated in illness (Ta-Nehisi Coates cited in Hoover, 2017, p. 235). The disparities in health outcomes for Indigenous peoples are a clear consequence of dispossession and an abrupt nutrition transition upon colonisation that continues in a climate where protecting a commodity, and corporate profits, is more important than protecting human health.

Accordingly, sugar remains a leader in 'dramatizing the tremendous power concealed in mass consumption' (Mintz, 1986, p. 185; Clarkson, 2020). In a pattern replicated across the colonised world, the Australian sugar industry was founded on slave labour. Between 1863 and 1904, over 62,000 South Sea Islanders were tricked and kidnapped or 'blackbirded' onto ships and transported to work on Queensland's sugar and cotton plantations (State Library of Queensland, 2019). By 1908, many had been deported under the 'White Australia' policy, a program to 'racially purify' the growing Australian nation and 'protect' white workers from the threat of cheap labour (Flanagan, Wilkie, & Iuliano, 2003). Currently our primary agricultural crop by volume, sugar uses more water than any other crop in Australia – 47% of the land that supports the crop is irrigated – and relies heavily on agrochemicals. The associated flow of inorganic nitrogen fertiliser, herbicides, pesticides and sediment into the Great Barrier Reef (GBR) catchment area, encompassing some 423,144 square kilometres and 35 river basins, has devastating impacts on fragile ecosystems. These include macro-algal growth, reduction of light available to seagrass and reefs, toxicity risk to inshore and coastal habitats, and threats from predators such as the crown-of-thorns starfish. Sugarcane farming and grazing are the primary sources of agricultural pollution in the GBR catchment area, according to a government Scientific Consensus Study published in 2017. Linking land condition, management practice standards and water quality outcomes, the report identifies 'a need to urgently implement more targeted and substantial effort to improve water quality in the Great Barrier Reef' (Waterhouse et al., 2017, p. 12). Climate change, however, remains the greatest threat to the reef, and the multispecies and livelihoods that depend on it.

For decades we have been joining the dots between food, health and environments. The social determinants of health determined by WHO recognise that the 'conditions in which people are born, grow, live, work and age…are shaped by the distribution of money, power and resources at global, national, and local levels' (WHO, 2020). The Sustainable Development Goals (SDGs) are designed to be a blueprint for systematically addressing these (UN, 2015). The SDGs demonstrate a growing awareness that climate change and environmental justice cannot be separated from other indicators of social, economic and physical well-being. They explicitly aim to 'ensure sustainable consumption and production patterns' (SDG 12), 'take urgent action on climate change and its effects' (SDG 13), 'conserve and sustainably use the oceans, seas and marine resources for sustainable development' (SDG 14) and 'protect, restore and promote sustainable use of territorial ecosystems, sustainably manage forests, combat desertification, and halt and reverse land degradation and halt biodiversity loss' (SDG 15). Yet they fail to tackle the way climate change will exacerbate inequity, and stop short of confronting the drivers of a food system that public health expert Professor Sharon Friel describes as *consumptagenic*. This is a system that 'encourages and rewards the exploitation of natural resources, excess production and hyperconsumerism, and which results in climate change and health inequities' (Friel, 2019, p. 136; see also Parker & Johnson, 2019; Banwell, Broom, Davies, & Dixon, 2012). Its 'backbone' is fossil-fuelled. While the SDGs observe the need to decouple economic growth from environmental degradation, they do not communicate with adequate force that 'the survival of our planet and the well-being of its people depends on our ability to reign in consumption and to change its character' (Friel, 2019, pp. 58–59). Further, they provide little in the way of a roadmap as to how this will be done.

A SHIFT IN WORLDVIEWS

Winona LaDuke, Ojibwe woman, environmentalist, feminist and Indigenous rights activist says food is 'one of the most powerful tools we can use to reshape our collective world-view' (cited in Fermanich, 2018). The lock-in that constrains change in our food systems is based on a particular worldview that, like any,

> ...once firmly entrenched, it tends to perpetuate a set of problems that are taken as natural and obvious. The possibilities of thought become calcified; the same questions and the same types of futile answer are repeated along the guidelines laid out by the grid that structures our thought.

> (Guignon cited in Mueller, 2017, p. 27)

Breaking out of this dominant paradigm is urgent given humanity's future depends on radical social change.

We might start by challenging the universality of the Anthropocene narrative. There is the 'post-nature'/'eco-modernist' vision of a high-tech future; an 'eco-catastrophist' narrative that draws on notions of local resilience in collapse; and the eco-Marxist version also known as the 'Capitalocene' (see Bonneuil & Fressoz, 2017; Holt-Giménez, 2017; Moore, 2017). The eco-feminist perspective – the Manthropocene? – blames the patriarchy for screwing us all. When it comes to food, Donna Haraway's (2015) notion of the Plantationocene is arguably the most appropriate. Derived from plantation agriculture, it highlights the impacts of extractive practices, monoculture development and exploitative labour structures underpinning socio-economic inequity and climate change. It provides a historic lens for analysing human-agented ecological change, and amplifies the differential impacts of these changes across different populations and regions. Beyond this, it highlights the prevalence of *cognitive injustice* (De Sousa Santos, 2007) which Gómez-Barris defines as a form of

> ...*constraining paternalism imposed on the Global South through colonising discourses and practices that continue to perceive these regions as purveyors of natural materials, and undervalue the heterogeneity of life embedded within local epistemes.*
>
> (Gómez-Barris, 2017, p. 99)

The Plantationocene calls attention to the ecological and economic legacies of empire, including the hierarchies and injustices that live on in our minds.

Foodways are also central to narratives that can sustain us. Especially valuable are lessons from ethnoterritorial social movements for whom 'sustainability involves the defence of an entire way of life, a mode of being-knowing-doing' (Escobar, 2017, p. 45). Anna Tsing (2015) suggests that these modes can emerge from the 'ruins of capitalism' where generative spaces foster the making of relations. Using the wild Matsutake mushroom as her guide in an uncertain world, Tsing explores precarious livelihoods and environments to help us 'appreciate the patchy unpredictability associated with our current condition' and 'reopen our imaginations' (Tsing, 2015, p. 5). Sharing stories of 'death, near-death and gratuitous life' along with diverse knowledge practices will not only help us face our current threats but 'offer our best hopes for precarious survival' (Tsing, 2015, p. 34).

Speaking and hearing these stories is a form of 'cultural, political, and intergenerational labour' required for 'taking care of the future' (DeLoughrey, 2019, p. 196). We can turn to them in critiquing and re-visioning the way we produce food. They are embedded in the cosmologies, imaginaries and practices of Indigenous peoples, small-scale farmers and land stewards globally. In agroecology, for example, a combination of agriculture and ecology that is capable of respecting diverse worldviews. Agroecology 'seeks a common matrix of dialogue between different realms, between traditional knowledge and Western science, putting them on the same level' (Altieri cited in Petrini, 2007, pp. 67–68). It does not impose 'one size fits all' solutions but promotes diverse site and scale appropriate ecological processes that enable ecosystems to self-regulate and carry out autonomously operations ranging from nutrient recycling to pest and disease control. Fundamental to agroecology as an alternative value system to the industrial model of agriculture is its role as a political strategy and praxis of change, activated through relations between co-producers in food communities (Martínez-Torres & Rosset, 2014; Méndez, Bacon, & Cohen, 2013; Mann, 2019; Holt-Giménez, 2002; Freire, 1970). Agroecological approaches recognise food as a network 'of men and women, of knowledge, of methods, of environments, of relations' (Petrini, 2007, p. 175). These days we readily acknowledge that we are increasingly enmeshed in digital networks but rarely stop to note that in ingesting food we become even more thoroughly 'materially entangled and implicated in a host of relationships' (Flood & Sloan, 2019, p. 15). In these complex relationalities, the tiniest critters in the soil, bees in a hive or polyps on a reef can teach us all 'to understand the parochialism of [our] ideas of individuals and collectives' (Haraway, 2019, p. 35). This is ecological reflexivity, and it is how we need to approach not just our food systems but our everyday lives.

On a personal level, we understandably look to our food choices to regain some kind of control over a world that is becoming increasingly uncertain and unstable. And our food choices do matter. Ethical principles such as transparency – knowing where and how food is produced; fairness – knowing our food does not impose costs on others; humanity – avoiding significant suffering of other species; social responsibility – ensuring workers have decent wages and working conditions; and needs – health and life above desires, may support the resilience of our food system (Singer, 2006). But do they have the capacity to transform it?

Focussing on personal consumption is worthy for those of us who can afford it, and vital for our health, but risks a reductionism that splits food out from the larger political project of creating equity and resilience in our food systems. Eating according to our ethics, even with the sincere goal of changing market practices, will fall short of achieving these goals. We all need to zoom out from our plates and see food as less of an object than a relationship, which forces us to address the commodification of human and non-human lives in the food system – and to think creatively about redesigning that system from the soil up in a spirit of radical hope.

ABOUT THIS BOOK

My approach embraces the biological, the economic, the political, the social and the cultural dimensions of how we produce and eat food. I aim to shift your perspective on food from that of a lifestyle choice to a political imperative, by using it as a lens through which to critically interrogate the ideologies, institutions and systems we have created, the basic structure of our society. I believe that while food provides ways of seeing the world, it also enables us to identify our agency in it. Most of all I hope to contribute to a conversation that 'fosters critical consciousness, a key prerequisite to effective social action' (Alexander, 2012, p. 15). In doing so, I draw on the stories and scholarship of researchers, activists and workers around the world.

This book is not going to tell you how to eat. Admittedly, it is a personal response to prevalent discourses that technological solutions and our personal 'purity politics' (Shotwell, 2016) are going to fix our food systems. As Ruby Tandoh (2019, p. 3) says, 'there is no single right way to eat, and you should be duly suspicious of anyone who tells you there is'. Pleasure is one of the highest principles we need to cling to in the struggle for food democracy (Carlson & Chappell, 2015). We need it if we are to overcome any apocalyptic rhetoric that shuts down the creative thinking essential to living well and creating messages that promote constructive and collective action. Furthermore, maintaining a 'fictitious separation between subsistence and pleasure' deprives millions of people of 'gastronomic dignity'. It denies the knowledge and practices that traditional cultures have 'built up and refined over centuries of adaptation to their land [which] continued to be plundered and expropriated from them' (Petrini, 2007, p. 40). It also discounts the value of diverse food cultures as sources of cultural transmission and insurrection (Nossiter, 2019).

My political ecology standpoint starts from the premise that environmental and ecological conditions and changes are the products of political processes. This includes the assumption that the costs and benefits of environmental change are unequally distributed between us and, more often than not, they reproduce and reinforce social and economic inequalities (Peet et al., 2011; Taylor, 2015). This understanding is built into the framing of the following questions, based on Jem

Bendell's 'Deep Adaptation Agenda' and applied to our food system:

Resilience – how do we preserve and maintain the foods and foodways (food practices, processes and behaviours) that we most value, those that will best aid our survival in a changing climate?

Relinquishment – what foodways do we need to let go of in order not to worsen human and planetary health?

Restoration – what can we repair, restore and recover in terms of ecosystems, attitudes and beliefs about food that will help us cope with the difficulties and losses that will face us?

Reconciliation – through our foodways, how do we make peace with our losses and mistakes to lessen our suffering, and that of all species, in the future?

To these I add a fifth 'R': *Reckoning.*

Reckoning is about owning histories of violence and exploitation; calling to account those who profit with impunity from foodways that continue to inflict hunger, poverty and despair. Where reckoning really counts is confronting, critiquing and resisting the continuation of a world food system dominated by wealth, markets and profits – one built on the dispossession of family and small-scale peasant farmers and the systematic persecution, stolen lands and genocides of Indigenous peoples (Ord, 2020).

The word *decolonisation* is often carelessly applied in a metaphoric way that promotes 'settler moves to innocence' rather than working towards actual decolonisation which means 'the repatriation of Indigenous land and life' (Tuck & Yang, 2012). I follow Indigenous scholars who note decolonisation 'is not a metaphor to be taken up within existing settler agendas for liberal, progressive, or radical politics that do not challenge ongoing settler colonialism' thereby keeping 'colonial relations and settler privilege intact'(Yazzie & Baldy, 2018, p. 4). Decolonising food systems means recognising food as a site of colonial struggle and breaking the lock-in presented by the capitalist food economy. It requires those of us advocating for food justice to listen to and be 'unsettled' by colonial histories to the extent that we revisit our basic assumptions regarding the land and our place in it (Mayes, 2018). 'First and foremost, decolonization must occur in our own minds' (Waziyatawin & Yellow Bird, 2005, p. 2). This includes speaking the 'hard truths' of colonialism, a process essential to generating the 'critical awareness that is necessary to heal from historical unresolved grief' (Lonetree, 2012, p. 6). It requires giving up what Sarah Maddison (2019) describes as the 'colonial fantasy' – the belief that colonialism is over.

Custodianship of land is central to Aboriginal cosmologies and cultural practices in my own country. The deliberate misunderstanding of this relationship was used to justify occupation but it remains an undeniable threat to the colonial fantasy by challenging the settler imaginary of complete sovereign authority over territory (Maddison, 2019, p. 76; see also Mayes, 2018). Reconciliation, a device transported with the settler colonies, has little power as a promise or a practice if it is not 'a reality' in the minds and actions of Aboriginal and non-Aboriginal Australians on a local level (Burnley cited in Maddison, 2019, p. 191). Wiradjuri woman Teela Reid (2020) says reconciliation has 'shielded the Australian state from its responsibility to rectify past wrongs through accountability and action'. Reckoning, on the other hand, 'requires everyday folks to bring about bold change'.

The shortcomings of reconciliation are confronted in the *Uluru Statement from the Heart* (2017), a national consensus position on Indigenous constitutional recognition, developed through 13 Regional Dialogues and a convention of 250 ATSIA delegates. For Reid it is

> *...a statement of truth about our history, and*
> *articulates a sophisticated roadmap to reckoning*
> *with our past through a series of reforms unique to*
> *the Australian context that aim to disrupt the system*
> *as we know it, beginning with establishing a First*
> *Nations Voice.*

(Reid, 2020)

Voice is central to disrupting food systems under neoliberal capitalism (Mann, 2019; Couldry, 2010; Butler, 2005). It is vital to dimensions of 'decolonial struggle' including resistance, critique, counterhegemonic mobilisation and tactical action (Smith, 2012). First Nations, agrarian movements and 'empty belly environmentalists' (Guha & Martínez-Alier, 1997) worldwide engage daily in these dynamic, material and intersectional contestations focused on life and land. Their work is multi-issue, 'seamed through with other economic and cultural causes as they experience environmental threat not as a planetary abstraction but as a set of inhabited risks' (Nixon, 2013, p. 4). They comprise what the Zapatista call the 'pluriverse' – *a world where many worlds fit* – an

> *...ethical and political practice of alterity that*
> *involves a deep concern for social justice, the radical*
> *equality of all beings, and non-hierarchy. It's about*
> *the difference that all marginalized and subaltern*
> *groups have to live with day in and day out, and that*
> *only privileged groups can afford to overlook as*
> *they act as if the entire world were, or should be, as*
> *they see it.*

(Escobar, 2017, p. xvi)

The challenge of producing and consuming food in a changing climate is much more than a conversation about what we grow and eat. It requires us to face historical and persisting inequities in our societies by engaging in the truth-telling demanded by the Uluru Statement – *'what is the truth and what does repair look like?'* (Davis, 2020). Only by reckoning with the politics of the industrial food system, and its deep colonial roots, can we unite and cultivate real change. The fact is our global food system is embedded in an economic system historically evolved from and *still committed* to the maximisation of economic growth. Reversing our consumerist mindset is just one element of building the food future we want (Wiedmann, Lenzen, Keyßer & Steinberger, 2020). This means looking beyond our foodways to establish a new vision of the future of society. We can live with uncertainty if it is accompanied by the understanding that 'to use the world well, to be able to stop wasting it and our time in it, we need to relearn our being in it' (Le Guin, 2019, p. 15).

STRUCTURE OF THE BOOK

Chapter 2: Food under Fossil Capitalism describes how the acceleration of food production and accumulation of capital for elites under the fossil fuel economy evolves to create highly unequal societies. The implications for the world's farmers, rural workers and the ecosystems they steward are analysed in this chapter through examples from Asia and Latin America.

Chapter 3: Framing the Future of Food explores how new and established 'protein producers' rehearse the existing industrial logic of increasing production and efficiency by developing meat and dairy substitutes. These 'victimless' products are marketed as more sustainable than livestock protein. Can they drive the cultural and behavioural changes needed to transform our food systems? Or will they further diminish biodiversity and resilience and the many forms of flourishing that relate to food?

Chapter 4: Changing Our Water Ways considers the complex challenges global warming presents to our waterways and the food contained in them, including longer droughts, rising seas and ocean acidification. Soaring levels of pollution from pole to pole and declining fish stocks are evidence that we are becoming increasingly reliant on aquatic species to supplement our diets. Is aquaculture the answer?

Chapter 5: Recovering Food Wisdom addresses how we can restore the health of all bodies and our ecosystems. It starts with a change in attitudes and the recovery of beliefs, norms and the nutritional wisdom that guides the traditional food practices of Indigenous people and regenerative farmers. Radical participatory methods and concepts like food sovereignty provide roadmaps for scaling cooperation and social capital across communities. They show us that the future, though unknown, is emergent.

Chapter 6: Resilience through Resistance turns to actions to reclaim our rightful places as food citizens, not just passive consumers. Critical, provocative, and rebellious, this resistance is being led by those most affected by the insults of the industrial food system. Leading through convening, they are amplifying the voices of those on the frontline. Joining in solidarity with them is vital if we are to 'get ahead of the game and begin creating cultures and societies of transition' (Yunkaporta, 2019, p. 81) that will sustain us in a changing climate.

2

FOOD UNDER FOSSIL CAPITALISM

THE MIRACLE OF MORE

Early in the pandemic, Australians were warned not be too smug about our food security. A net exporter of food with strong supplies of local fresh produce (75% of total food supply) we are in a good place. Until the tin cans – along with the plastic packaging, the chemical fertiliser and the food additives – run out. We export beans and legumes but can't can them for our own supermarket shelves without access to global supply chains because our last tinplate steel mill closed in 2006 (Simons, 2020b).

A masterpiece of human design, the tin can represents modernity in our food system. In the short film *The Miracle of the Can* (1956), it is presented as a 'shining beacon of economic growth based on technological progress and the extraction of limitless natural resources, resulting in ever-increasing production..."much as we have today, our promise of the future is ever more and more" ' (cited in Flood & Sloan, 2019, p. 14). Its story is one of colonial empire building. Malaysia was the largest tin producer in the world until competition from Latin America grew and the market slumped in the 1980s. Today, in a rising market, the Indonesian island Bangka hosts the world's largest off-shore mining fleet which extracts 3.5 million tonnes of tin ore from the ocean floor per month. Most of it now goes to the electronics industry to build our smart phones. In the state-owned Tanjung Pesona mine, there are no safety rules or environmental safeguards but mining 'death metal' is better paid than fishing so there is no shortage of available labour – men, women and children. The processing of the ore leaves stagnant and polluted pools of water that host malarial mosquitos, and leaves toxic, barren soil where lush coastal forest and fertile farmland once protected the beaches from erosion (Ifansasti, 2012).

Scientific and industrial developments that make food more durable and transportable appear to provide all the advantages we need to ensure a stable global food supply. These innovations mask the complex political geographies behind our consumption; ideologies and forms of 'militarised production' (Gómez-Barris, 2017, p. 4) in extractive zones like Bangka. The out-sourcing of toxic industries appears justified if it produces the objects central to our very existence as consumers (where are we without our phones?). Upon exhaustion at one site, raw materials are appropriated from another part of an ever-diminishing, shifting 'periphery' and transported to an expanding 'core' where bottomless markets will find uses for them.

The slow, attritional violence inflicted on communities like Bangka 'overspill clear boundaries in time and space [and] are marked above all by displacements – temporal, geographical, rhetorical, and technological'. For example, the impacts of toxic irradiation on the Marshall Islands in nuclear tests in the 1950s and Mururoa Atoll in the 60s continue in the form of congenital birth defects. These displacements 'simplify violence and underestimate, in advance and in retrospect, the human and environmental costs'; they 'smooth the way for amnesia' (Nixon, 2013, p. 7). Marshall Island diplomats are now among the most strident and influential voices at climate change summits, and have called out major carbon emitters for 'cultural genocide' (DeLoughrey, 2019, p. 4).

Our carbon economy, in which growth is predicated on the increasing consumption of non-renewable fuels and the continued emission of carbon dioxide, relies on the extraction of raw materials in regions with cheap and plentiful pools of labour and low-regulatory controls. This is a form of 'environmental burden shifting' – the out-sourcing of the environmental impacts of our consumption preferences to 'less-visible countries and communities elsewhere, including to those along global supply chains' (UNDP, 2019, pp. 176–177). Burden shifting goes beyond GHG production to waste generation, meat consumption and water use. It facilitates the production of monetary surplus through exploitation of labour that transforms raw materials into consumer goods. It is in this relationship that 'fossil capital' emerges as the root cause of our 'warming condition' (Malm, 2016).

THE ORIGINS OF FOSSIL CAPITAL

In pre-industrial agriculture, the organic economy was based on human and animal power and photosynthesis. Yield was restricted by the supply of land. Fossil fuels eliminated these barriers. A new mineral-based economy opened up enormous opportunities for growth, fuelling capital relations within which competition is the rule of law. The Great Divergence of Europe from the rest of the world was propelled in equal measure by the discovery of coal, self-interest and material consumption, despite evidence of destructive impacts. 'Clean air acts' date back to the late 18th century in response to the health and environmental impacts of sewers, mines and polluting factories. Farmers pointed to chemical works and gas lighting to explain crop diseases (in 1852, two Belgian farmers died protesting against the local soda factory for the failure of their potato harvest). The politicisation of climate was especially strong in France in the 1820s with bad harvests being blamed on the Revolution, the division of communal forests and their rapid demolition by the new consumptive class, the bourgeoisie. It was noted that fumes respected no borders in Europe and therefore governments needed to cooperate. This awareness was countered by the argument that chemical plants helped 'regulate the overall composition of the atmosphere' while in France coal was presented as a 'green energy' to mitigate shrinking forests (Bonneuil & Fressoz, 2017, p. 175). Charles Babbage observed, in 1832, that steam engines were 'constantly increasing the atmosphere by large quantities of carbonic acid and other gases noxious to animal life'. While acknowledging that 'the means by which nature decomposes these elements, or reconverts them into a solid form, are not sufficiently known', he qualified this concern with the idea that an immense, recycling nature 'acting over vast spaces, and unlimited by time' would cope (cited in Bonneuil & Fressoz, 2017, p. 205).

As industrial production and urbanisation increased so did perceptions of the 'metabolic rift', an irreparable rupture between humanity and the rest of nature. Here Karl Marx critiqued a capitalist agriculture that imposed its laws on the environment and broke the bonds between town and country by disrupting the metabolic interaction between humans and nature. He wrote 'it prevents the return to the soil of its constituent elements consumed by man in the form of food and clothing; hence it hinders the operation of the eternal natural condition for the lasting fertility of the soil' (Marx, 1976, pp. 637–8). While the Spanish were importing sodium nitrate as mineral fertiliser from Chile in the 1830s, the declining fertility of English soil was supplemented by relentless imports of guano from Peru and

Bolivia enabled by the "coolie" trade – large numbers of Chinese bonded labourers working under conditions so appalling that guards had to be posted to prevent suicides (Foster & Clark, 2020, pp. 25–26). This led to a characterisation of the British Empire as 'Vampire-like. It clings to the throat of Europe, one could even say of the whole world, sucking its best blood' by Justin von Liebig (cited in Bonneuil & Fressoz, 2017, p. 187). He described Britain's evolving industrial agricultural system as the "robbery economy" (*Raubwirtschaft*), including the theft of human lives in his critique (Foster & Clark, 2020, p. 13).

Liebig's discovery that adding sulphuric acid to ground bones made calcium phosphates more readily available to plants led to the development of single super phosphate, one of the first examples of chemical agronomy. It was, however, the ground-breaking synthesis of ammonia by German duo Fritz Haber and Carl Bosch in 1910 that marks the starting point of large-scale production, in the form of nitrogen fertiliser. This has enabled us to increase grain production fourfold since 1950, and human population threefold, by converting hydrocarbons into carbohydrates, and thus food. But we have overwhelmed the Earth's capacity to absorb nitrogen. Run-off is a major contributor to the Gulf of Mexico's hypoxic or dead zone, measured at 8,776 square miles in 2017. Excess nutrients stimulate algal growth which consumes all the oxygen that would otherwise be available to fish and shrimp, destroying the food chain. Increased nutrient loading is supported by US farm policy which subsidises the industrial corn and soy fields that are the source of most of the 1.15 million metric tons of nitrogen pollution that flowed into the Gulf *in 2016 alone* – 170% more than the BP oil spill of 2010 of 670, 800 metric tons (Mighty Earth, 2017; Thill, 2017; Philpott, 2020). Like all artificial fertilisers and pesticides based on petroleum products, our widespread application of nitrogen fertiliser is a major contributor to GHG emissions.

These technological developments drove the 20th century's Great Acceleration, an era of dramatic socio-economic growth in the wealthy Organisation of Economic Co-operation and Development (OECD) countries. Aggregates of development data for this period mask extreme inequities and vast differences in national prosperity and well-being globally, and the development of diverse economies. Industrialised countries came to rely less on agriculture as a source of GDP (an average of 2%) than poor countries (30–50%) making the latter especially vulnerable to shocks in the world food economy. In 2011, a huge 91 per cent of Guinea-Bissau's, 80 per cent of Malawi's, and 70 per cent of Nicaragua's exports were agricultural products. New Zealand stands out as an exception among the WEIRDs, at 50 per cent (Clapp, 2016, p. 10).

From the mid-20th century on, we witnessed not only a widening gap between rich and poor countries but fundamental shifts in the state of Earth systems. This period marks the dawn of the Anthropocene for Earth System scientists (Steffen, Broadgate, Deutsch, Gaffney, & Ludwig, 2015). They warn that human activity has pushed us beyond several of the critical boundaries that enable life to flourish on Earth, an 'ecological ceiling' that includes climate change, ocean acidification, chemical pollution, nitrogen and phosphorus loading, freshwater withdrawals, land conversion, biodiversity loss, air pollution and ozone layer depletion.

AWARE, BUT NOT ALARMED

The concept of boundaries, and our tendency to exceed them, is not new. In 200 BC, Quintus Septimius Tertullianus noted 'one thing is sure: the earth is more cultivated and developed now than ever before: there is more farming but fewer forests, swamps are drying up and cities are springing up on an unprecedented scale. We have become a burden to our planet. Resources are becoming scarce and soon nature will no longer be able to satisfy our needs' (cited in Pringle, 2003, p. 84). Over 2000 years later *Limits to Growth: A Report for the Club of Rome's Project on the Predicament of Mankind* (1972) proclaimed 'if the present growth trends in world population, industrialisation, pollution, food production, and resource depletion continue unchanged, the limits to growth on this planet will be reached sometime within the next one hundred years'. The authors of the report suggested that 'it is possible to alter these growth trends and to establish a condition of ecological and economic stability that is sustainable far into the future...so that the basic material needs of each person on earth are satisfied and each person has an equal opportunity to realise his individual human potential' (Meadows, Meadows, Randers, & Behrens, 1972, p. 24).

On the 20th anniversary of the report, the updated *Beyond the Limits* (1992) argued that we had 'overshot' planetary limits. The causes of overshoot are growth, acceleration and change, exceeding a limit or barrier over which is not safe to pass. The final, critical element is a 'delay or mistake in the perceptions and the responses that try to keep the system within its limits...arising from inattention, faulty data, a false theory about how the system responds, deliberate efforts to mislead, or from momentum that prevents the system from being stopped quickly'.

The original and so prescient assessment of *Limits to Growth*, based on a system dynamics approach to hard data analysis in a computational model called World3, was criticised as underestimating the power of technology and the 'adaptive resilience' of the free market. But a 30-year update justifies its scepticism, with the attendant explanation:

> *One reason technology and markets are unlikely to prevent overshoot and collapse is that technology and markets are merely tools to serve goals of society as a whole. If society's implicit goals are to exploit nature, enrich the elites, and ignore the long term, then society will develop the technologies and markets that destroy the environment, widen the gap between rich and poor, and optimize short term gain.* In short, society develops technologies and markets that hasten a collapse instead of preventing it.
>
> (Meadows, Randers, & Meadows, 2004, pp. 8–9)

It's not like we weren't warned. As Bill McKibben reminds us in *Falter* (2019), climate change has been a public issue for 30 years. Prior to this, popular concepts like Spaceship Earth (Fuller, 1969) and Gaia (Lovelock & Margulis, 1974) urged citizens and policy-makers to accept that Earth is somewhat like the squeeze ball – when we exert pressure in one part of the system, it bulges elsewhere in unpredictable ways. In her landmark *Diet for a Small Planet*, Frances Moore Lappé (1971) told Americans that their grain-fed meat-centred foodways were akin to driving a gas-guzzling Cadillac, often. The counter-narrative of 'sustainable development' was successfully deployed to allay all these fears. Continuous economic growth could be achieved through careful management of planetary resources based on linear and reversible calculations such as maximum sustainable yield; in fisheries, for example, catches rose from 20 million tonnes in 1950 to 80 million in 1970, leading to massive declines in reserves. Increasing GDP was pitched as a conservation strategy – once wealth trickles-down and people escape from poverty they will have the time, resources and attention to care for their environments.

Free-market environmentalism emerged from the economic logic that marketable rights to pollute exist, and that actors should negotiate among themselves to ensure the 'polluter pays'. There followed a suite of solutions to environmental challenges based on market instruments: water and biodiversity markets, fishing and groundwater quotas and 'ecosystemic services'. Land is appropriated for activities rewarded with carbon and biodiversity credits that transform nature into 'a source of wealth extraction for big business and global finance' (Seufert, 2020, p. 9). Justified as 'conservation' but more accurately described as 'full-belly environmentalism', many projects preserve pristine, natural environments by limiting access of local and Indigenous people with the effect of destroying their livelihoods, cultures and autonomy (Guha & Martínez-Alier, 1997). This 'instrumentalization of the whole world and the Earth, this limitless devaluation of everything given' (Arendt, 1958, p. 157) had many implications, including the end of limits to growth.

In late capitalism, world-systems are increasingly measured in flows of matter and energy which have contributed to the calculation of ecological footprints incorporating the notion of 'ghost acres' – territory appropriated through importing and consuming products from other regions (Pomeranz, 2000). These spatial constructions attempt to measure the unequal exchange and dispossession that drive the production of 'cheap' food (McMichael, 2005; Patel & Moore, 2018). Yet they fail to adequately capture the social misery and environmental costs of a precarious world food economy made all the more volatile by climate change.

MAKING SENSE OF THE WORLD FOOD ECONOMY

Oversupply of food has been a constant companion of hunger for decades. This contradiction can only be understood within the context of global political and economic relations which have opened up 'middle spaces' along food supply chains where the 'norms, practices, and rules that govern the world food economy are shaped by the very forces that are leading to its expansion' (Clapp, 2016, p. 7). Harriet Friedmann and Philip McMichael's systematic formulation of food regimes (Friedmann & McMichael, 1989) is useful in unpacking what McMichael describes as 'a foundational divide between environmentally catastrophic agro-industrialisation and alternative, agroecological practices that is coming to a head now as we face a historic threshold governed by peak oil, peak soil, climate change, and malnutrition' (McMichael, 2009, p. 141).

As discussed in Chapter 1, the forced transplantation of people and crops through colonisation, slavery and plantation agriculture laid the foundations of the modern food system. It led to 'new hybrid species, and a global homogenisation of the Earth's biota…a radical reorganisation of life…without geological precedent' (Crosby cited in DeLoughrey, 2019, p. 24). Populations in Asia, Europe and Africa doubled with the support of food exports from the Americas, fuelling development and industrialisation. This diaspora, literally 'spore and seed', includes pollen maize, discovered in European marine records of the early 1600s, signalling the transfer of crops between the Americas and Europe and, arguably, the start of the Plantationocene. This period also registers a drop in carbon dioxide because of reduced farming in the Americas; the direct result of 50 million Indigenous deaths caused by small pox (Devlin, 2015).

Plantation agriculture provided the foundations for the colonial food regime (1870–1930). Under British imperialism, economic globalisation was shaped by merchants, bankers and elite citizens generating fortunes in mining, textiles and, later, finance. The slave trade was still underway, supporting the sugar and cotton industries. Tropical imports – coffee from Brazil, bananas from Central America, peanuts from Senegal – along with grains and livestock produced in settler colonies, were sent back to Europe. A great reversal in food security occurred in this period. Famines disappeared from Western Europe while millions died in China and India. Local populations and ecosystems in the new colonies were decimated by the deforestation of the Caribbean for sugar plantations, the polluting of rivers by silver mining in Mexico and Peru, plus the virtual extinction of the beaver, bison and bowhead whale in North America. Trans-continental infrastructure supported the colonial food regime and anchored the periphery to the European economy through canals, railways, ships, docks and telegraph lines. The United Fruit Company, for example, created a built-to-order railway system in Guatemala that was

> ...ideal if you were a banana trying to get to a shipping port, but rather irrelevant if you were a person trying to get from one place to another in your own country.

(Gálvez, 2018, p. 97)

Integration led to dependencies exploited by the advanced economies in the post-war food regime (1950s–70s). After the Second World War, US hegemony was promoted by an emergent agrofood complex in which farm supports, protectionist policies and technological innovations in crop and livestock breeding, mechanisation and chemical inputs contributed to the production of 'historically unprecedented' food surpluses (Friedmann & McMichael, 1989; Patel & de Wit, 2020, p. 177). Seeking new markets in Europe and then the developing world, the US set a trend for Australia, Canada and later Europe, when it recovered from the War. Surpluses provided domestic benefits through food relief and school lunch programs – and soon became a source of overconsumption through the 'supersizing' of consumer portions and the creation of new ingredients like high fructose corn syrup (HFCS). Excess food was also deployed as foreign food aid, which eliminates storage costs, reduces downward pressure on domestic prices and is a means of opening new markets. Notably, this served political objectives in the Cold War period as a counter to communism. 'Communism has no greater ally than hunger; and democracy and freedom has no greater ally than an abundance of food' according to US Congressman Hubert Humphrey in 1953 (cited in Clapp, 2016, p. 53). The Green Revolution was described in the same terms by William Gaud, Administrator for the US Agency for International Development (USAID) who noted, in 1968, it was 'not a violent red revolution like that of the Soviets, nor a white revolution like that of the Shah of Iran' (Clapp, 2016, p. 36). The Green Revolution enabled increased yields of wheat, maize and rice in countries like India but came with strings attached; the provision of aid was conditional upon 'full adoption and support' of the technologies, locking in a new industrial agricultural model in regions like the Punjab (Clapp, 2016, p. 39). This 'short-tether' approach exacerbated inequality between rich and poor farmers within countries, leading to increased tensions and violence.

Into the corporate food regime (1980–) heightened integration of trade within and between emerging agrifood transnational corporations (TNCs) furthered the globalisation of the food economy and the legitimisation of 'a wholesale conversion of the global South into a "world farm"' (McMichael, 2009, p. 287; see also McMichael, 2005). The Green Revolution was supplemented by a Gene Revolution featuring the application of biotechnology to genetically engineer plants. Rather than invest in this research, the US government relied on the private sector which continues to profit from it (Clapp, 2016, p. 44). Meanwhile intensive meat production stepped-up in the United States, and Europe accelerated production of surplus grains for export (Weis, 2007).

Global supply chains connected national farm sectors and generated transnational divisions of labour in agriculture. Latin America and Asia experienced agricultural growth rates beyond population expansion; in Africa, food produced per capita declined in good part because of the lack of investment in small family farms (Hawkes, Lobstein & Lang, 2012). Use of heavy agricultural machinery, industrial pesticides and fertilisers rose exponentially with long term damaging impacts to the environment including the depletion and pollution of soils, waterways and entire ecosystems, and massive land clearing for crops and pasture. High energy and transportation costs increased the price of food locally and also generated new international markets for biofuels. In Latin America, green deserts of sugar cane, a cash crop for ethanol production, displaced family farmers (Rundgren, 2016, p. 104). US and EU biofuel policies diverted cereals, especially corn, from food to biofuel stock. This was a significant factor in the 140% increase in food prices experienced between 2002 and February 2008 which may have been a blip on the radar for WEIRD consumers but was devastating for the world's poor who spend upward of 50%–80% of their total income on food (Clapp, 2016).

Food price crises stem from more than a misalignment of supply and demand; macroeconomic conditions, energy policies (including biofuels), financial speculation and trade are all implicated. These factors can be traced back to the rise of corporatist states as part of an agenda of economic modernisation and capitalist development throughout the 20th century. This agenda was facilitated by two global policy developments. Firstly, structural adjustment programs (SAPs) gave loans to countries in the Global South on the condition that they reform national economies by opening markets, deregulating and privatising industries, removing regulatory controls and devaluing currencies. Secondly, an emerging free trade agenda created agreements (FTAs) promoting the global integration of markets and the creation of accommodating conditions for the private sector through mechanisms including intellectual property.

Trade, 'the seeking out of methods of buying as cheaply as possible what is needed, and of selling most advantageously what can be produced for sale' (Brillat-Savarin, 1949, p. 61), is intrinsic to foodways. It is an age-old source of revenue and means of exchange between communities. The neoliberal trade regime denies 'the utilitarian and individualistic spirit of the trader, which has profited him for centuries' (Petrini, 2007, p. 229). Instead, rules for imports and exports are set by international development agencies and states 'on behalf of agrifood monopolies [and] adjudicated in part by lobbyists in proceedings where trade agreements are hammered out beyond democratic view' (Patel & de Wit, 2020, p. 179).

The efforts of the World Trade Organisation (WTO) to liberalise agricultural trade through the Agreement on Agriculture (AoA) by reducing farm subsidies in rich countries and opening global markets to 'free' trade instead produced 'an uneven playing field' (Clapp, 2016, p. 15). The entire regime is now in disarray, with stagnating trade talks, the decline of the United States, Brexit, and the rising power of China contributing to the instability of 'a state system already compromised by capitalist power, deterritorialization and deepening political polarizations' (McMichael, 2020, p. 29). All of which is complicated by our warming condition and the pandemic.

FREE TRADE UNLEASHED

Mexico provides an excellent example of how agricultural trade liberalisation has not only created dependencies based on 'unbalanced rules' but also 'fostered ecological crisis through the encouragement of large-scale export-oriented agriculture' (Clapp, 2016, p. 94). Christian Parenti describes Mexican fisheries, in particular, as 'the story of climate change expressing itself through the political economic realities of neoliberalism' (2011, p. 186).

Following the Revolution (1910–1920) Article 27 of the Mexican Constitution dictated that the state owned all resources, ranging from land to seas and petroleum. The principle underpinning this document was land reform to bring about a 'more equitable distribution of public wealth...to create new agricultural centres, with necessary land and waters; to encourage agriculture in general and to prevent the destruction of natural resources' (Mexican Constitution, 1917). By 1949, 23% of land was collectively owned in the *ejido* communal land management system (up from 1.6% at the time of the Revolution) and by 1960, 20,000 *ejidos* comprising 2 million members worked almost half all cultivated land. Prime fishing stocks were reserved for small-scale fishers and state-sponsored cooperatives. The latter were granted rights to the most significant inshore marine and shellfish areas while subsistence fishers claimed the remainder. Processing, packing and marketing were controlled by the state enterprise *Productos Pesqueros Mexicanos*. The industry was embedded in layers of regulation defined by economic nationalism, a protectionist profile that was soon challenged by the 'Globalization Project' (McMichael, 2020).

Lured to participate in the sphere of international commerce, Mexico became 'a laboratory for policies that frame foreign investment as the path to prosperity and development' (Gálvez, 2018, p. 86). To spur on the economy, 'progressive' reforms and policies of importsubstitution industrialisation (ISI) were instituted. These enabled Mexico, with many other Latin American countries, to create 'limited labour-capital compacts' in which the state owned some industries and controlled others – 'a semisocialist set of interventions' that represented 'an alliance for profit' with business (Parenti, 2011, p. 190). These measures were the undoing of nationally controlled assets and protections, enabling Mexican capitalists to develop monopolies and cartels. Elite factions promoting conservative economic policy grew. Rising inflation, increasing public debt and social unrest culminated in the massacre of hundreds of students at a protest at La Plaza de Tres Culturas at Tlatelolco in October 1968. In response, a neo-populist program of political and social reforms was funded by oil income, which grew by almost two-thirds between 1979 and 1980.

The Mexican economy steadily liberalised and joined in trade agreements with Canada and the US. State enterprises like *Productos Pesqueros Mexicanos* were privatised. The country continued to borrow against its 'petrodollars', diversify areas of economic growth, and repay its international debts, all of which came with strings attached. Austerity and increased exports were among the conditions imposed by the International Monetary Fund (IMF) and the World Bank. This led to an oversupply of commodities including timber, ores, grains, sugar, coffee and oil. The economy stagnated while farmers and fishers endured the early deprivations of climate change in the form of severe droughts. The collapse of the peso in 1982 led Mexico to enter into further compromise, selling state-owned companies including sugar mills, textile processing and power plants. In 1989, foreign ownership of up to 50% of industries was allowed, and subsidies to small fishers were reduced. The impacts included closure of more than 75% of publicly owned businesses, the decline of the co-operatives, and rising poverty in fishing communities. With the withdrawal of the state, corruption and poaching increased – by 2010, over 12,000 unregulated fishing boats were estimated to operate in the Sea of Cortez. Mexico's share of catch now accounts for just 10%, the rest going to the US, Canada and Japan. Coupled with increasingly unpredictable extreme weather, poor economic policies are creating climate refugees who are fleeing the coast to find work in cities and across the US border.

The ultimate expression of Mexico's liberalised economy is the North American Free Trade Agreement (NAFTA), finalised on 1 January 1994. Free trade and climate change have eviscerated the Mexican 'corn-tortilla value chain'. Once protected by government policy, hundreds of Indigenous varieties of maize have been erased or adulterated by genetically modified crops. Between the 1930s and early 2000s Mexico lost 80 per cent of traditional varieties (Clapp, 2016, p. 52). In 2009–10 the worst drought in decades decimated crops of maize, wheat, beans and sorghum. Desertification is leading to the migration of 600,000 to 700,000 people annually. Once largely self-sufficient, today Mexico imports 42% of its food, has a health crisis of obesity and diabetes, and possesses poverty rates over 50%. The 'death of the family farm' has created a food economy where 'local green markets have been replaced with supermarkets overflowing with processed food and soda'. Foodies in Manhattan celebrate handmade tortillas made of landrace corn 'while Mexicans have become the world's top consumers of instant noodles' (Gálvez, 2018, p. 10).

They are the casualties of a 'globalization/diet nexus' (Hawkes, 2006) where food trade, global sourcing, direct foreign investment, food advertising, supermarkets, global agribusiness and TNCs converge. These dominant firms have become 'too big to feed humanity sustainably, too big to operate on equitable terms with other food system actors and too big to drive the types of innovation we need' (IPES, 2017, p. 5). Their immense pricing power gives them massive advantages in financial markets, actively reducing competition within the agricultural industry through mergers and acquisitions (Clapp, 2014). They also exert influence by lobbying governments to establish favourable trade and investment rules, and engage in private 'standard-setting that shapes the types and characteristics of products provided by suppliers' (Clapp, 2016, p. 98). Examples include the Roundtable on Sustainable Palm Oil (RSPO) which hosts the top four commodity trading firms and the Global Roundtable on Sustainable Beef (GRSB). Set up to reduce the environmental harms of beef production, the latter is comprised of the biggest names in the industry including Walmart, Cargill, JBS and McDonald's (Winders & Ransom, 2019). Embedded deeply in the market, and in the carbon ideologies that drive the production of cheap meat, their decisions have profound impacts on local producers (Du Toit, 2001).

The rise of private standards demonstrates how states have stepped back from their regulatory roles while the leaders of agrifood TNCs position themselves as proactive and progressive in tackling sustainability issues. This is not unlike giving the inmates keys to the asylum. The technology of ethics that underpins corporate responsibility schemes operates according to the neoliberal logics of profit, unlike fair trade and organic standards, which typically include a wider range of stakeholders, including NGOs. Food retail is governed by environmental, economic, and social sustainability criteria devised by mixed stakeholder groups like GlobalGAP which sets standards for over 150,000 suppliers in over 100 countries (Clapp, 2016, p. 123). Their results are mixed, bringing benefits like improvements in food safety to consumers while disadvantaging small farmers and bringing down the wages of farm workers, particularly those in the Global South who do not have the resources to meet more stringent demands. Modern slavery persists: a 2018 study of tea and cocoa plantations in India and Ghana revealed that ethical certification schemes fail to prevent exploitation and forced labour (Nolan & Boersma, 2019). Domestic chains like 'Colesworth' (Coles and Woolworths) in Australia squeeze farmer incomes and erode their ability to choose what to grow, how to grow it and for whom; an independent review of the Grocery Code of

Conduct found them guilty of 'unfair tactics and arbitrary decision-making with little regard for the potential damage to suppliers' business' (Samuel, 2018, p.17; Mann, 2019). The size and influence of the modern supermarket, with its capacity to rezone cities, set wages, channel capital and enact a form of soft power has been identified as 'Wal-mart capitalism'. In the US, the decline of trade union power and weak government regulation has given big retailers 'more power than any other entity to "legislate" key components of American social and industrial policy' (Lichtenstein, 2005, p. 22).

While the activities of TNCs and major retailers have been forensically examined by food systems analysts, the complex entanglements of global finance and food are less visible. Since the 1990s derivatives sold by banks, investment brokers and grain traders have focused on food commodities, sometimes bundled with non-food products. Agricultural commodity futures markets played a role in the 2007–2008 food price hikes, and a subsequent record high in 2011; the Commodity Food Price Index rose by almost 75% between 2006 and the end of 2010, causing wheat prices to surge 56% (Parenti, 2011, p. 229). There is disagreement on the extent of the impact of fluctuations of financial markets in a 'perfect storm' that included poor harvests, growing demand for high-protein diets, and the diversion of crops for biofuels (Clapp, 2016, p. 134; Bello & Baviera, 2010). However grain trader Cargill openly attributed an 86 per cent increase in annual profits in the first quarter of 2008 to its commodity futures trading business, and Bunge and Archer Daniels Midland (ADM) posted similar results (Clapp, 2016, p. 154). There is little doubt that the increasing incidence of extreme weather events and erratic seasonality triggered by climate change will play a major role in speculation over crops, impacting the most vulnerable peoples' access to food.

Volatility in food prices is exacerbated by the commodification of the bio-economy, whereby crops such as corn, soy and palm oil become what McMichael calls 'fungible investments' virtually interchangeable as feed, fuel or fertiliser. In these transformations, exchange value 'erases' actual value in the 'ultimate fetishization of agriculture, as an input-output process geared to indiscriminate production of commodities for profit' (McMichael, 2013, p. 132). Commodities like corn and soybeans are today reduced into component parts for recomposition according to flavour, texture, colour and other criteria. Agrifood TNCs control these supply chains, sourcing, at the lowest cost possible, undifferentiated, commodified foodstuffs such as refined grains, sugars, fats and salts obtained from multiple producers (Vivero-Pol, 2017). These global dynamics have contributed to the development trope of the 'zone', a once-rural place subjected to a 'disremembering' driven by 'corporate interests pursuing their appetite for ever more accumulation' (Berger, 2007, p. 116). This is illustrated clearly in the case of the most valuable, internationally traded 'forest risk' (Trase, 2018b) commodity driving climate emissions: the soybean.

SPOTLIGHT ON SOY

Originally from Southeast Asia, the soy plant is a nitrogen-fixing legume that is well-suited to temperate climates. China is the world's biggest importer and leader in both soybean meal and oil production. The crop is vital to China's pork industry, dominated by domestic agribusiness firms known as 'Dragon Heads'. These are unlike any other commercial enterprise in the global food system in terms of their role and responsibility, tasked with 'opening up new markets, innovating in science and technology, driving farm households, and advancing regional economic development' (Schneider, 2019, p. 87). In 2011, there were 110,000 officially designated by the state.

Argentina, Brazil and Paraguay now produce almost 50% of the world's soybeans on a land area that has increased forty-fold since 1970 to an area larger than Spain. Demand for the crop spiked in the 1970s with the collapse of the Peruvian anchovy fishery. It escalated into the 1990s, resulting in widespread deforestation of the Amazon biome. By 2019, soybean exports from Brazil amounted to around US$26 billion, over 78% going to China as high-protein animal feed and refined cooking oil (Statista, 2020). Over 36 million hectares of land in Brazil are devoted to soybean production which has expanded by about 1 million hectares a year on average over the past decade (Meyer & Schipani, 2019). The trade is dominated by a handful of megacompanies: in 2016, Bunge, Cargill, ADM, COFCO (China's largest food processing company), Louis Dreyfus and Amaggi accounted for 57% of all soy exports from Brazil.

The soybean expansion frontier has shifted from the Amazon to the Cerrado, a vast tropical savanna ecoregion in Midwestern and Southeast Brazil which neighbours the Amazon. It embraces portions of the states of Maranhão, Tocantins, Piauí and Bahia, collectively referred to as Matopiba (Lima, Da Silva, Rausch, Gibbs, & Johann, 2019). This migration was propelled by the establishment of the 2006 *Soy Moratorium* (SoyM), which prohibited major soybean traders from purchasing soy grown on Amazonian lands deforested after 2008. Aligned with government and NGO campaigns to reduce deforestation, coupled with transparent and rigorous systems of monitoring and compliance, SoyM is considered a success (Gibbs et al., 2015). However, soybean cultivation is still responsible for Amazonian deforestation through the displacement of pastures for beef, another major competitor for land in the region (Barona, Ramankutty, Hyman, & Coomes, 2010, p. 8).

The Cerrado biome is home to more than 4,800 unique species of plants and vertebrates – over 5% of global biodiversity. It is a crucial source of evapotranspiration, feeding up to 12 hydrological regions in Brazil. Exempt from SoyM and the Brazilian Forest Code, which requires the retention of 80% of forest on a property, the Cerrado has lost 46% of its land area to soy (Lima et al., 2019). Local farmers report the decline of small tributaries, streams and aquifers which they rely on for their water supply, and attribute this to intensive irrigation (Trase, 2018a). Deforestation over the past 30 years has contributed to an 8.4% decline in rainfall in the region, which is further threatened by extended droughts with global warming (Dijkhorst, 2018). Coupled with logging and mining, deforestation of the Cerrado is contributing to the disruption of the rainfall regime in the Midwest and Southeast regions of Brazil. Given the rapid rates of change, these previously lush and fertile biomes do not have the capacity to regenerate (SlowFood, 2020). Local farmers are displaced as land ownership becomes more concentrated; in 2010, more than 50% of soybean cultivation was controlled by just 3% of producers. The rural exodus caused by this model of production means many of those driven out of the countryside are condemned to urban poverty.

Conflicts in the Cerrado continue a long history of resistance against Brazil's industrial agribusiness expansion by the landless peoples' movement, MST (*Movimento dos Trablhadores Rurais Sem Terra*) (Fernandes, 2015). In a 2017 action in the municipality of Correntina, hundreds of locals destroyed the facilities of a major agribusiness farm used to grow grains and vegetables to protest its high water usage – up to 176 million litres per day from the Arrojado River. This is enough to supply the whole town for more than a month. Attributing water shortages to these big farms, one local claims 'they are getting worse...they say they bring jobs, but [they] destroy our river' (Prager & Milhorance, 2018). In 2017, *The Cerrado Manifesto* (CM) was declared with the aim of curbing deforestation and conversion of native vegetation. It called on soy and meat producers to 'disassociate' their supply chains from recently converted areas. A coalition of over 70 companies including McDonald's, Unilever and Walmart, and 50 investors signed (Cerrado Manifesto, 2017).

Yet far-right, populist leader and proud anti-environmentalist Jair Bolsonaro, elected Brazil's president on 1 January 2019, is changing the direction of environmental policy. Moving the deforestation control sector of the Environment Ministry to Agriculture and abolishing climate change from the political agenda, he has stated that 'not a single centimetre' of land will be demarcated for Indigenous peoples and that protected areas and Indigenous lands should be open to agriculture and mining. In short, 'the actions of President Bolsonaro and his ministers favour expansion of monoculture plantations and cattle ranching in Amazonia' (Ferrante & Fearnside, 2019, p. 262). November 2019 saw the appearance of a campaign to end SoyM and a backtrack on CM. Bartolomeu Braz Pereira, president of Brazilian soy producer association *Aprosoja Brasil*, declared, 'They came with plans to expand the moratorium to the Cerrado and we said we're going to overturn it in the Amazon'. He cited support from Bolsonaro, who agrees that SoyM was an imposition on Brazilian farmers the government was committed to contesting (Samora, 2019).

The entire Southern Cone, spanning Brazil, Argentina, Uruguay, Paraguay and Bolivia, is known colloquially as 'The Republic of Soy' after Syngenta Corporation labelled it such in a 2003 campaign in national daily newspapers *Clarín* and *La Nación*. This corporate-driven model of territorial organisation is 'a regionalized, hi-tech re-enactment of the extractive, commodity-dependent model of economic growth historically known to Latin America' dating back to the monetisation of European economies by Peruvian gold and Bolivian silver in the 15th century (Turzi, 2011, p. 65). The Soybean Republic has huge 'strategic importance' as 'importing countries intensify their quest for food resources' (p. 66).

Food security and soy production 'run on different tracks' notes agronomist Carlos Todelo (cited in Frayssinet, 2015). But soy is a hard habit to break for those who do farm it. Genetically modified soy is considered a less risky crop, is comparatively loosely regulated and much cheaper to produce than maize, beef or dairy products. If farmers can't see advantages to planting other crops, and if soy brings in the best tax revenue for the state, that lock-in is here to stay.

While the impacts of soy exports devastate the Brazilian countryside and its communities, they also play a huge role in the restructuring of China's peasant economy. The Dragon Head model is supporting corporate power and consolidation, shifting control to companies and away from small-holder farmers. This aligns with the discourse in both countries that peasants are 'a problem for which further capitalist industrialisation is the only and inevitable solution' (Schneider, 2019, p. 92). Meanwhile 'diseases of affluence' are rising with Chinese pork consumption as a growing middle-class quadruple meat consumption from what it was in 1980 to 65 kg per person, per year. Twenty-three per cent of boys and 14% of girls in China are overweight or obese (Schneider, 2019, p. 94). Food safety scares over growth-promoting hormones, antibiotic-resistant diseases, epidemics of swine fever and eutrophic waterways including a dead zone in the East China Sea caused by fertiliser runoff are unintended consequences of China's pork boom.

THE GLOBAL LAND RUSH

Large-scale foreign land acquisitions are the latest iteration of the financialisation of food under fossil capitalism. Post-colonial 'land grabs' date back to Green Revolution which impoverished millions of the rural poor and forced them from their land (Wolford, Borras, Hall, Scoones, & White, 2013). Today's land grabs frequently fly under the radar with the consolidation of farms driven by global financial firms and foreign pension funds including Swedish national pension fund, Blackstone and Harvard University (GRAIN, 2015). Investigations of farmland acquisitions in Brazil by New York–based College Retirement Equity Fund (TIAA-CREF) have revealed links to local businessmen known to be engaged in violent practices of *grilagem*, the leveraging of political connections and false documents to claim title over public lands and forests, justifying the eviction of local people. The *chapadas*, fertile areas throughout the Cerrado with good access to water, are particular targets as they facilitate highly mechanised models of production employed in broadacre farming. Local peasant, Indigenous, and Afro-Brazilian or *Quilombola* communities are devastated as their sources of food, animal feed, firewood and medicines are eradicated, and they are driven to find work in hyper-urbanised cities, local diamond mines or sugarcane plantations.

Pension funds are attracted to farmland in depressed and volatile markets as farmland possesses 'good fundamentals' in an insecure food future (Holt-Giménez, 2017). The commodification of land and formation of 'agri-finance capital' is aided by the application of information and communication technologies (ICTs) 'to develop standardised metrics to assess the potential for earning profit from farmland, and create agile business forms that can increase farmland liquidity' (Spadotto, Sawelijew, Frederico, & Pitta, 2020, p. 2). Targeted countries, where land prices are relatively low, include Australia, Sudan, Uruguay, Russia and Zambia. The Zambian government is partnering with a blockchain company to assist with land registration and titling to unlock mineral reserves that are currently inaccessible under informal land governance

(Mousseau, Currier, Fraser, & Green, 2016). Who will control these data, and how they will be used, is of great concern to the ETC Group who note that technological innovation in robotics, sensors, gene editing and 'fintech' (blockchain and cryptocurrencies) open up huge potential for the generation of Big Data that will be owned by corporations, further disempowering the rural poor (Abrol, 2019).

In response to growing concerns over the expropriation of public land, the World Bank and International Fund for Agricultural Development (IFAD), the UN Conference on Trade and Development (UNCTAD) and the FAO created the *Principles for Responsible Agricultural Investment that Respects Rights, Livelihoods and Resources* (PRAI) (FAO, 2014). Regrettably, these soft instruments give the benefit of the doubt to land users who promise infrastructure and services that 'do no harm' to local environments. Arable lands and productive forests and fisheries are being diverted from small-holders to commercial, extractive industries and monocultural production, including biofuels, violating the right to food of local populations whose informed and prior consent is not sought. In countries such as Cambodia and the Philippines, 'agricultural investment' projects involve the trade of loans for produce and land for biofuel feedstock with wealthy but land-poor countries such as Kuwait and Qatar (Bello, 2009). Weak states with poorly regulated national systems for land title and registration, and particularly those experiencing conflict and high rates of hunger and poverty such as Liberia and Uganda, are ripe targets.

Critics of PRAI, including those involved in the Civil Society Mechanism (CSM) of the World Committee for Food Security (WCFS), demand the implementation of genuine land reform to ensure equitable allocation of resources along with support for agroecological, participatory methods of food production and governance of local and regional markets, and the protection of community-based food and farming systems from corporate and state control (FIAN International, 2010; GRAIN, 2010, 2015; PCFS, 2013; TNI, 2015a). The Peoples' Committee for Food Sovereignty (PCFS) is especially scathing of the neoliberal elements of the framework of the PRAI, namely 'its facilitation of the monopoly and control of big corporations and wealthy states over the global food system' (PCFS, 2013, p. 1). The secret to resolving land grabs is addressing this impunity and supporting small to medium farms to diversify their production. But this does not form part of the business plan of a fossil-fuel driven food system run on the principle of 'get big, or get out'.

POST-PANDEMIC THINKING

This chapter has provided a brief overview of how policies focused on the privatisation, deregulation and liberalisation of national economies have historically pressured poorer countries to adopt measures that support the expansion and concentration of corporate wealth and power. The impacts of global markets ricochet across oceans within a system designed specifically for the production of cheap food. In this accounting, we value products and services without recognising the ecological function from which they are derived 'resulting in a perverse incentive to degrade the Earth's ecosystems' (cited in Schwartz, 2013, p. 203).

Vital biomes such as the Cerrado and the Amazon become battlefields between cattle ranchers, soy growers and Indigenous peoples in competition for wild environments that once provided an abundance of food. COVID-19 is evidence of how 'expansion of the agricultural frontier and the incursions to search and capture wild specimens for commercialisation' bring people and animals into dangerously close contact, 'creating the ideal conditions for the incubation of the most diverse viruses, allowing the overflow of diseases from one species to another' (SlowFood, 2020). The pandemic highlights the 'massifying and standardizing' food production model as the source of the ills of our food system and is a prime opportunity to 'rethink food and farmers' models'.

According to the peasant farmers movement La Vía Campesina (2020) COVID-19 has 'exposed the vulnerability of the current globalised food system dominated by industrial agriculture, and the dangers it poses to all life forms', suggesting that 'we should learn from the crisis and invest in building local, resilient and diverse food systems'.

Where do we start?

3

FRAMING THE FUTURE OF FOOD

WHAT DOES INNOVATION TASTE LIKE?

Eating is a far more complicated behaviour than it seems. The food we eat entangles us in communities, economies and politics, and implicates us in life-sustaining, symbiotic alliances with other species.

> *When we eat we taste the world around us, entering into a relationship with plants, animals and even the microbes we collaborate with to create some of our tastiest, most celebrated foods.*
>
> (Flood & Sloan, 2019, p. 121)

Equally complex are the narratives that shape how and what we eat. These are increasingly sophisticated and well-targeted, particularly by food producers and retailers whose careful attention to product design, packaging and branding is often designed to cultivate more 'caring relationships with the "distant strangers" behind our everyday eating' (Sexton, Garnett & Lorimer, 2019, p. 48). Many food companies claim to help us make better or healthier food choices. Relations between consumers and producers are conditioned by food labels and advertising containing icons and devices which

> ... *reflect a socio-political environment in which consumption is deemed to be an appropriate, if not a preeminent, field through which to exert influence over the ethics of the entire food system.*
>
> (Evans & Miele, 2017, p. 233)

Social and embodied food practices play a significant role in shaping our eating routines. National cuisines speak to our religious and political heritage and also serve as testament to changing climatic and ecological conditions (Giggs, 2020). Food is deeply politicised, 'interlarded with relationships of power and privilege'. Our everyday food choices reveal and reproduce relations and divisions of cultural, social and economic capital within a 'broad discourse where ideologies of consumerism and citizenship shape different understandings of how food relates to equity, social justice, and sustainability' (Baumann & Johnston, 2010, p. 128). The rising popularity of vegetarian and vegan diets in WEIRD societies is an 'expression of expanding spheres of concern' with human health, animal suffering and the environmental impacts of 'containment agriculture' including effluent run-off, antibiotic discharge and land clearing for feed crops (Giggs, 2020, p. 217).

Negotiating our food-shaped world, or 'sitopia', boils down to the question of how to live (Steel, 2020). Our food cultures are so embedded and intrinsic to our identities that we are actually what we eat (Brillat-Savarin, 1949), and the impacts of our food practices ripple throughout the food chain. We are 'agents of selection, whereby the genes, genomes and ingredients that are propagated are the ones you prefer to eat...every human eater slowly reformulates the planet as they consume it' (Flood & Sloan, 2019, p. 121). With this comes responsibility and the weight of lives. How we strike a balance between our own nourishment, and the interests of others, is neither a solitary, one-off decision, nor is it resolved in a vacuum (Giggs, 2020, p. 218). But can the strictest vegan protect the farmworker, or the bee for that matter, from pesticide exposure? It seems that to eat, no matter how we do it, is to kill.

Or is it? The new heroes of the Anthropocene narrative include the food scientists and biotechnologists developing novel food products in labs. Like their predecessors, the architects of genetic modification and the Green Revolution, they promise to resolve hunger. We are vulnerable to the compelling solutions offered by these and other 'experts', including food manufacturers, for whom the omnivore's dilemma is not so much a dilemma as an opportunity.

> *It is very much in the interests of the food industry to exacerbate our anxieties about what to eat, the better to then assuage them with new products.*
>
> (Pollan, 2006, p. 5)

Marion Nestle is skeptical we can trust an industry that promised to provide solutions to poverty and hunger in the Global South but 'instead concentrated on far more profitable insect- and herbicide-resistant first-world crops' leading to widespread misery and environmental harms. 'Trust requires fulfilled promises' (cited in Krimsky, 2019, pp. x–xi). Heroes and villains abound in the fraught field of novel food products, where innovation is pitched against risk in a battle including activists, farmers, scientists and consumer groups.

A new clutch of entrepreneurs suggest technology is the answer. But we might well ask – what was the question? (Price, 1979).

ENGINEERING NUTRITION

In 2019, one of *Time* magazine's '100 Best Inventions' was Soylent, a drink that offers a complete meal in every bottle. It looks, and tastes, like a protein shake or the milk left over in your cereal. It is made of a plant-based protein derived from soy which is relatively easy to produce and has high bioavailability, which means it is rapidly absorbed into the body. Marketed as an 'easy, affordable, delicious solution to help you avoid food voids' – which are, apparently, 'the place[s] where you're stuck eating something you'll likely regret or when you don't eat at all' – Soylent claims it is helping to mitigate the damaging effects of food production on the planet through application of genetically modified organisms (GMOs).

> [H]ere at Soylent, we want to change the way people look at food. That's why we are pro-science and pro-GMO. GMOs are a safe, economic option for sustainable food production, they cut down on food waste, time spent growing food, and resources used.
>
> (Solyent, 2019)

Could the solution to the challenges facing our food systems be as simple as adopting meal replacements designed and produced in the laboratory? This is indeed the case according to think-tank RethinkX, which proclaims that 'we are on the cusp of the fastest, deepest, most consequential disruption of agriculture in history' (Tubb & Seba, 2019, p. 6). RethinkX believes proteins will be five times cheaper by 2030, before 'ultimately approaching the cost of sugar'. They will be superior – 'nutritious, healthier, better tasting and more convenient, with almost unimaginable variety'. Animal-agriculture will be replaced by the Food-as-Software model, in which 'foods are engineered by scientists at a molecular level and uploaded to databases that can be accessed by food designers anywhere in the world'. This will allegedly result in 'a far more distributed, localised food-production system that is more stable and resilient than the one it replaces' – largely because it will not be subject to the impacts of weather, drought, disease or even seasonality (p. 7). GHG emissions will be dramatically decreased: those from cattle will drop 60% by 2030, and net emissions across all livestock will decline by 45% by 2030, and 65% by 2035, according to RethinkX's modelling. Demand for oil will 'disappear'; net water consumption will be reduced by 35% by 2030, and 60% by 2035 (p. 8). Basically, RethinkX claims that by 2030 modern food products will be better in every way, and cost less than half what we currently spend on the animal-based products they replace.

This is a dire forecast for the dominant industrial livestock production model which is at its limit in terms of scale, reach and efficiency – and is reportedly one of the most significant factors in global warming. According to the FAO, the meat and dairy industries contribute 14.5% of anthropogenic GHG emissions, 20% of which are accounted for in fossil fuel-based transport (Gerber et al., 2013). Animal products provide 37% of protein and 18% of calories for humans yet are responsible for 83% of farmland and 56–58% of food-related emissions (Poore & Nemecek, 2018). Raising animals for food con-sumes a third of the planet's fresh water, and occupies up to half of the Earth's habitable land. Ninety-four percent of mammal biomass, excluding humans, is livestock – they outweigh wild mammals by a factor of 15 to 1. Agriculture and aquaculture are listed as a direct threat to 24,000 of the 28,000 species vulnerable to the 'sixth' extinction (Kolbert, 2014; Ritchie & Roser, 2020).

Should the world make a dietary shift away from animals, we would reduce food-related GHGs by 49% (Poore & Nemecek, 2018); and we would save a lot of land. The directors of the 2014 documentary *Cowspiracy* claim that feeding a vegan for a year requires just one-sixth of an acre of land; a vegetarian who eats dairy and eggs three times that; the average American diet including meat, dairy and eggs 18 times more – a whole three acres (Kuhn & Andersen, 2015, p. 61). Beyond alerting us to the environmental impacts of meat production, the critical message of *Cowspiracy* is that we are 'kept blind…by design' to the cruelties and environmental crimes of industrial meat production by those who profit from it. Timothy Pachirat, who worked for six months in a Nebraska slaughterhouse to inform his book *Every Twelve Seconds: Industrialized Slaughter and the Politics of Sight*, sums up a vital truth that extends beyond the 'sites of production' of industrial food:

> *In all political processes where the unacceptable must be rendered acceptable, where the morally and physically disgusting must be made digestible, fabrication departments – literal and allegorical – perform a dual work of construction and manufacture and of framing, forgery and the invention of legends and lies.*
>
> (Pachirat, 2013, p. 32)

It's clear from all of this that there is a strong argument for a shift in diets, particularly in WEIRD societies where meat consumption is highest. Plant-based diets are on the rise in developed economies like the US and the UK, where 5% of the population now identify as vegan (Kaufman, 2018), yet per capita global consumption of animal products has more than doubled since 1980. Americans consume twice the recommended intake of protein despite well-publicised negative health impacts (Foer, 2019). In Asia, meat consumption is rising 4% per year; milk and dairy 2–3%. Livestock is the fastest growing sector of agriculture, particularly for the new middle-class in emerging economies like China, which accounts for 28% of the world's meat consumption, primarily pork. Over the past 10 years, US milk production has increased by 13% due to high market prices. However, Americans are drinking less milk; 149 pounds of milk/capita in 2017, compared to 247 pounds in 1975, according to USDA data. As a result, the country has a massive 1.4 billion-pound cheese surplus (Raphelson, 2019). Global demand for animal protein may double again by 2050 (Friel, 2019).

These contradictions are classic symptoms of what Tony Weis refers to as the industrial grain-oilseed-livestock complex and the 'vector of meatification' (Weis, 2013, p. 8). He links the uneven meatification of diets directly to climate change and presents it as a major marker of global inequality whereby

> ... the climb up the 'animal protein ladder' is part and parcel of the climb up the 'development ladder', and patterns of rising meat consumption at the national scale have been very tightly linked to patterns of rising affluence, with industrialised countries consuming meat at vastly higher levels and the world's poorest regions at the bottom of the meat consumption spectrum.
>
> (Weis, 2013, p. 71)

Given this complexity, is a shift to engineered nutrition and plant-based meat replacements a realistic or viable response to the ills of our consumptagenic food system? Or is it just another market opportunity, which distracts us from better solutions for our warming condition – solutions that also address persisting inequalities, injustices and vulnerabilities within our food system and contribute to ecosystem health?

CULTIVATING TASTES FOR TECHNOLOGY

The power of marketing to influence modern food culture and the uptake of new technologies is profound. Good storytelling underpins the dairy-alternative market, projected to be worth US$14 billion and reporting growth of more than 60% between 2013 and 2018. Goodfoods dairy-free cheeses, manufactured from a cashew nut base combined with fermented pine nuts, nutritional yeast and spices, claim to produce 99% fewer GHG emissions and use less water, land and energy compared to dairy-based counterparts (Southey, 2019b). The US dairy industry, which saw giant Dean Foods bankrupt in 2019, is under pressure and responding vigorously by opposing the labelling of non-dairy products including plant-based milks.

In the stories constructed to promote the personal and planetary health benefits of novel plant-based foods 'narrative and technology reinforce one another in powerful feedback loops, each contributing their share to revealing the world in a particular manner' (Mueller, 2017, p. 27). Milk has moved from the 'original superfood', replete with calcium, protein and vitamins, to 'scary dairy', catastrophic for both animals and the environment. Plant milks are at the vanguard of the clean-eating craze promoted by millennial influencers on

Instagram who have achieved what animal rights and vegan activists had tried to do for decades. 'Suddenly, giving up milk wasn't just a health issue…it was part of living your best and most beautiful life' (Franklin-Wallis, 2019). Today, brands like Nutty Bruce, Oatly and Rebel Kitchen (creator of Mylk) jump on trends for 'activated' nuts, antioxidants and blood typing. Their language is perfectly pitched to the 'post-milk generation'. Oatly welcomes the 'future oatmilk drinker' with self-congratulatory banter:

> *We just know that despite the fact that there are a lot*
> *of plant-based options out there to add to your coffee*
> *or morning cereal, the combo of oatsome*
> *deliciousness and what our products can do for you*
> *is rather challenging to find elsewhere.*

Oatly's foamy 'barista edition' is a huge hit in hipster enclaves like Brooklyn in New York and Shoreditch in London. Other plant-based milks rely on the same rural aesthetic as traditional dairy – pastoral idylls and friendly farmers – to 'emphasize both the vision of a tamed, benevolent nature and a faith in technological innovation to resolve agricultural struggles' (Bladow, 2015, p. 9). Some brands, like Califia Farms almond milk, also draw on the health qualities of dairy with direct comparisons such as 'contains 50% more calcium than milk' suggesting that plant-based products are interchangeable with their animal-based counterparts. In doing so plant-based brands adeptly exploit the shift to 'functional nutritionism', which embraces techniques to reengineer the nutritional profile of foods and offers new opportunities for food companies to market the 'precise health benefits' of their products (Scrinis, 2013, p. 158). Almond milks make up two-thirds of plant milk sold in the US (Franklin-Wallis, 2019). Since 1998, when Blue Diamond released Almond Breeze, the first US commercial almond milk, lyrical brand names and health claims have belied the environmental impacts of almonds; chiefly their high water requirements. The nut was brought to the US by Spanish missionaries but not commercially grown until the mid-1800s when French varieties were widely cultivated by growers associations and cooperatives. Their water footprint quickly became apparent – one Californian almond averages 12 litres (Fulton, Norton, & Shilling, 2019). In Central Valley, source of 80% of the world's almond supply, the crop has grown from 500,000 acres in 2000 to cover an area the size of Delaware. It produces one million tonnes of almonds annually. Faced with a deepening water crisis, California has increased water use efficiency through more sophisticated irrigation systems and soil management practices (Bladow, 2015). In Australia, the world's second largest supplier, the nut uses triple the water required for wheat or feed grain – a minimum of 8.5–10 mega litres per hectare throughout the growing season from October to April (Hannam, 2019). Planting along perennial river systems ensures a relatively reliable supply of water but increasingly extreme dry weather periods are bringing almonds into competition with horticulture, dairy, rice and cotton (Aither, 2019; Simpson, 2016; Day, 2019). Despite misgivings over water security 15,000 hectares of new trees have been planted in the vital Murray-Darling River Basin, a 50% increase since 2016, leading growers to call for a moratorium on new plantations and a stocktake of water resources (Davies, 2019).

Almond trees require highly fertile soil that has low salinity levels and low clay content. Therefore agro-chemical inputs are commonly applied to crops. They are also grown in monocultures that receive chemicals lethal to bees. Fungicides and pesticides that tackle alternaria, anthracnose, brown rot, green fruit rot, hull rot, leaf blight, scab, shot hole and rust combine to create toxic environments for honeybees and their larvae. While fungicides are not highly toxic, they synergise with miticide treatments used by beekeepers and create considerable challenges. Pesticides affect bee navigation, social learning, sperm viability, gut microflora, foraging behaviour, worker productivity and pollen collection efficiency. Colony collapse disorder (CCD), first identified in 2006, has been traced back to pesticides including glyphosate. The loss of natural habitat driven by monocultures deprives bees of the safe haven they need from pesticides and prevents access to diverse nesting sites and florals (Noble, 2014). Large industrial farms strip the orchard ground bare to more effectively treat for insects and fungi, eroding biodiversity. Almond pollination requires more hives than any other crops. Apples, America's second largest pollination crop, use only one-tenth the number of bees. California's almond mega-farms require so many bees that commercial keepers send hives across the country. Senior scientist for the Center for Biological Diversity, Nate Donley, describes this practice as 'sending the bees to war. Many don't come back' (cited in McGivney, 2020). Tracheal mite infestations, Africanised 'killer' bees and parasitic mites affect around 30% or more bees per year. Fifty billion – one-third of commercial US bee colonies – were killed during the winter of 2018–2019.

Farmers are now breeding almond varieties that require fewer hives per acre to pollinate. A 'Bee Better' certification program was launched in 2017 to help identify almond products where biodiversity has been introduced to naturally control pests and nourish honey bees. Small-scale sustainable almond farming that allows greater biodiversity and doesn't use pesticides yields a smaller crop but provides healthier conditions for bees (McGivney, 2020).

The potential of other crops to satisfy plant-based milk palates is being explored. Swebol Biotech and a team of Bolivian scientists recently developed Quiny, a milk alternative using royal white quinoa from the Andes, available in two formats: ready to drink and an instant powder. It contains no added sugar. The ready-to-drink version has a 'clean label' ingredient list – ingredients we understand, free of e-numbers. It is expected to have great potential in South America's plant-milk market, particularly because quinoa is native to the region and already widely grown (Michail, 2019). Yet the impact of the 'quinoa moment', like that of landrace corn in Mexico, has a complex dynamics. Its global popularity as a superfood has led Andean quinoa farmers to replace the grain in their own diets with cheaper and less nutritious rice and noodles. Sudden demand has also led to sustainability issues. The 'inevitable bust' that has followed is not visible to the WEIRD consumer or canny investor who might not challenge the logic, or the ethics, of commercialising Indigenous foods under the guise of climate-proofing our diets (McDonell, 2018).

> *Eating a plant-based diet in Europe defeats the purpose if this means eating avocados from Mexico or quinoa from Peru or Bolivia, or consuming meat alternatives that are ultra-processed foods, wrapped in plastic, and produced by major corporations like Unilever.*
>
> (Vanessa Álvarez González cited in Morena, 2020, p. 54)

PLANT-BASED BUSINESS-AS-USUAL

The story of plant-based milks exposes the danger of fetishising novel foods. It demonstrates how we need to go beyond analysis of the nutritional makeup of products to ingredients to consider how these were grown, with what impact and for whose benefit (Goodyear, 2015). The same scrutiny is due to the entire plant-based protein market, predicted to be worth over US$480 billion by 2024. The industry anticipates massive growth based on shifting consumer attitudes, namely a desire for 'more sustainable options' (Fortune, 2019b). Nestlé reports that 87% of Americans already include plant-based protein in their diets. Established 'protein producers' like Tyson Foods, Purdue, Smithfield and Cargill are also targeting the 'flexitarians' who maintain a vegetarian diet with the occasional serve of meat or fish.

Public scepticism taints plant-based brands entering into business relationships with major meat companies. In particular, Tyson Foods' heavy investment in alternative-meat technologies raises concerns that the company is not only exploiting new market niches but greenwashing an unethical business model. The company has a poor environmental pollution record, including a (paltry) US$2 million fine for violating the Clean Water Act by knowingly dumping waste from a chicken processing plant into the Missouri River (Kaufman, 2018). It has been accused of paying lip-service to the humane treatment of farmed animals with the set up of a 'strategic and superficial' independent Farm Animal Well-Being Advisory Panel which 'makes for good publicity but offers little hope for real improvements in animal well-being' (Wells, 2013). Upon accepting capital investment from Tyson Foods in 2017, Beyond Meat founder and CEO Ethan Brown was told he had 'blood on his hands'. Brown defended the deal, telling Bloomberg his approach is pragmatic as 'the people at Tyson know how to move the needle' (Piper, 2019).

Whether that needle is moving people to more sustainable diets, or Tyson into more profits, becomes clearer upon investigation. Noel White, CEO of Tyson Foods, has declared

> ... for us, this is about "and" – not "or". We remain firmly committed to our growing traditional meat business and expect to be a market leader in

> *alternative protein, which is experiencing double-*
> *digit growth and could someday be a billion-dollar*
> *business for our company.*
>
> (Tyson, 2019a)

The company has no plans to stop its stock-in-trade and will sell plant-based products alongside their meat products (Yaffe-Bellany, 2019). In June 2019, only a few months after divesting from Beyond Meat, Tyson Foods unveiled its 'Raised and Rooted' brand featuring plant-based nuggets and a blended burger made with Angus beef and pea protein isolate which it claims has fewer calories and less saturated fat than the plant-based burgers sold by competitors. In September 2019, it ventured into the seafood-replacement market with New Wave Foods, which planned to have a shrimp alternative ready for food service operators in 2020 (Tyson, 2019b). Compassion Over Killing, an animal rights organisation based in Washington D.C., is highly critical of Tyson's duplicity noting 'it's clear that Tyson is beginning to take notice of the growing demand for ethical options but refuses to sever ties with the barbaric meat industry'(Compassion Over Killing, 2019). The company is not 'self-disrupting' by investing in plant-based meat products but rather 'making bets on where the commodities will go next' (Newman, 2020).

Novel food start-ups are based on the Silicon Valley model, where protein synthesis replaces software as the core technology. This attracted early celebrity investors including Bill Gates. Market leader Impossible Foods is colloquially known as 'Impossible Patents' with 14 registered to date, and 100 pending for chicken and fish proxies (Itzkan, 2020). Chief Executive Pat Brown declares he is on a mission to eliminate all animal products from the global food supply by 2035, claiming that the Impossible Burger spared the equivalent of

81,000 tonnes of GHG emissions and 900 million gallons of water in 2018. He sees big opportunities to promote plant-based pork in China where demand for meat exceeds by a factor of four what the country can produce on its own land (Bloomberg News, 2019).

The United Nations is on-board, endorsing Impossible Foods as

> ... *one of a handful of companies proving that sales growth and consumer interest can coincide, at a massive scale in harmony with planetary health, by displacing large-scale, historic sources of resource degradation.*

> (UNFCCC, 2019)

Now partnered with Burger King in the US (serving the 'Impossible Whopper') and Hungry Jack's in Australia, Impossible Food's not-so-secret ingredient is heme – specifically leghemoglobin – a vegetable version of the substance in animal muscle that gives the burger the aroma, taste and mouthfeel of meat. It is effectively genetically altered, fermented yeast. Aside from the moral panic that persists over GMOs, the fermentation processes used to grow the yeast rely on feedstock in the form of high inputs of sugar. Given the footprint of King Sugar, as discussed in Chapter 1, this is arguably displacing threats – by shifting the impact of grazing livestock for meat in temperate climates to already stressed, and particularly climate shock prone, tropical regions (Newman, 2019).

Scientists are calling out the false promise that ultra-processed plant products are the most climate-friendly alternative to giving up meat. Upon launching the 'Big Vegan' in Germany, Nestlé tweeted that it was meeting consumer demand for food 'with less impact on the environment' with a burger made of soy and wheat protein, beetroots, carrots and bell peppers. Oxford University researcher Marco Springmann and colleagues determined that the carbon footprint of this and many other processed plant-based products falls between chicken and beef (Lucas, 2019). They suggest that real alternatives to animal protein already exist in the form of the bean patty, which has a fifth of the footprint of plant-based meat substitutes.

The truly novel foods riding the plant-based push sidestep the issues associated with soy and pea proteins, two of the most common ingredients in plant-based products. The soybean, as discussed in Chapter 2, is derived from vast tracts of monocultures that contribute to deforestation, particularly in Latin America, and allergies to it are common, while pea protein has a strong flavour which requires additional processing to mask, driving up the price and environmental footprint of the final product. Companies creating genuinely disruptive and breakthrough technologies include Israeli start-up Food for the World (FFW), focusing on yeast-based products, and Perfect Day (formerly Muufri) which supplies ADM with synthetic ingredients produced through fermentation to replace milk proteins. Should the latter reach their goal of producing a satisfactory replacement for dairy fat, they might also displace the palm and coconut oils used in many plant-based products, thereby helping these products live up to sustainability claims (Hughes, 2020). FoodSolutions Team's NaSu burger is made entirely from food production side-streams, primarily from juicing and edible oil production, with the main ingredients being carrot fibre, beetroot fibre, flaxseed flour and pumpkin seed flour (Askew, 2019). Finnish company Solar Foods, whose slogan is 'food out of thin air', is claiming its Solein powder produces five times fewer GHG emissions than plant-based foods, 100 times less than beef, far less water and, of course, much less land. Aiming to produce sufficient protein for two million meals per year, it aspires to feed 9 billion by 2050.

There is likely application for many of these new products to become part of a balanced diet, particularly where basic nutrition is lacking. The same goes for insects. A staple in Asia, South America, Africa and parts of Europe, species like beetles, caterpillars, bees, ants, crickets, grasshoppers and locusts are now promoted as a 'viable food group' in the West with high protein content and outstanding production efficiency compared to conventional industrialised foods. Europe and the US have fast growing edible-insect markets, expected to exceed US$522 million in value by 2023 (Kim, Yong, Kim, Kim, & Choi, 2019). They clash with WEIRD eating habits, but are they any less attractive than cell-based, lab-cultured meat?

NEW FOOD CULTURE(S)

The world's first lab-grown burger was cooked and eaten at a London news conference in 2013. Consisting of tens of billions of lab-grown cells and costing US$330,000 in research and development (R&D), it was described as 'close to meat, but not that juicy'. People for the Ethical Treatment of Animals (PETA) responded positively, proclaiming that lab-grown meat 'will spell the end of lorries full of cows and chickens, abattoirs and factory farming. It will reduce carbon emissions, conserve water and make the food supply safer' (BBC, 2013). The idea of lab-grown meat dates back to, and probably beyond, the 1930s when Winston Churchill said

> ... *fifty years hence we shall escape the absurdity of growing a whole chicken in order to eat the breast or wing by growing these parts separately under a suitable medium.*

> (Churchill, 1932)

The idea has caught on in Australia where lab-grown meat business VOW has been awarded AU$25,000 to develop lab-grown kangaroo meat products. VOW co-founder Tim Noakesmith says

> ... *growing meat sustainably from stem cells will have a fraction of the footprint of traditional livestock farming in terms of land use and water use and there is no need for culling animals.*

He plans to build a full-scale factory in Western Sydney that will supply tonnes of cell-based meat to domestic and international markets (Fortune, 2019a).

Its proponents argue that in vitro meat production potentially represents a 'revolution of human-, animal- and environment-friendly meat'. Yet data reveal that 80% of Americans would not eat lab-grown meat (Sharma, Thind & Kaur, 2015). This might have less to do with taste than the way it is produced. Cultured meat originates from the stem cells of living animals which are reproduced in a bioreactor. They are immersed in a growth medium such as glucose, foetal bovine serum, chicken embryo extract or synthetic amino acids which is the source of nutrients and energy for muscle cell production. It is cultivated in an open concrete pond and, once harvested, is sterilised and hydrolysed to break down the cells. Expensive inputs, including the supplements cell-based meat require to become nutritionally equivalent to real meat (including iron and Vitamin B12), and heavy reliance on the industrial energy required to manufacture and purify culture media make the process less palatable (Eenennaam, 2019). However, its supporters argue cultured meat production emits substantially fewer GHG emissions and requires only a fraction of land and water compared to conventionally produced meat in Europe. The energy requirements of cultured meat production are lower than those of beef, sheep and pork but higher than poultry. Studies report substantially lower nutrient losses to waterways compared to conventionally produced meat because wastewaters from cyanobacteria production can be more efficiently controlled compared to run-offs from agricultural fields (Tuomisto & Teixeira de Mattos, 2011).

Atmospheric modelling approaches that compare cultured meat processes with different modes of beef production reveal that 'despite the bold claims and superior carbon dioxide equivalent footprints, cultured meat is not necessarily a more sustainable alternative' (Lynch & Pierrehumbert, 2019). Why? Because cultured meat production mainly emits carbon dioxide, which has a millennial lifespan and changes the atmosphere slowly, while conventional beef production mainly emits methane, which has a lifespan of 12 years but changes the atmosphere very quickly. So the long-term impact of cultured meat production is dramatically worse than beef as its 'warming legacy' persists even if production declines or ceases. Cultured meat is thus not more climate-friendly than beef as its relative impacts depend on the source of energy generation and the livestock production system compared.

Basically, until energy generation is decarbonised, cultured meat is likely to remain unsustainable from a cost and energy perspective, a quandary that also faces hydroponic, vertical farms like Freight Farms, Boston (Coggan, 2020). These technologies are likely to reinforce the dominance of multinational food companies which possess the capital to invest in the R&D required to not only get products beyond the laboratory but also figure out how best to sell them to consumers once they're on the shelves.

WINNING HEARTS, MINDS AND STOMACHS

Those marketing novel food products need to acquire a
deep understanding of how we react to messaging on
conscious, rational and subconscious levels. A 2019 survey
indicates consumers rely on a balanced combination of
'scientifically supported facts and benefits' (Shoup, 2019). Yet
there is evidence that personal benefit trumps corporate
overreach, environmental sustainability and animal welfare
for many;

> ... at the end of the day, it's about is there a benefit
> for me personally, is it better or healthier for me,
> does it mean I can avoid hormones or antibiotics,
> does it have a great taste or flavour, does it provide
> more variety in my diet? Is it more convenient?
>
> (Demeritt cited in Watson, 2019b)

Whatever their motivations, younger generations are more
open to and quicker to embrace plant-based proteins and lab-
grown meats than older generations. Major barriers for wider
uptake continue to be cost and negative taste perception. But
food manufacturers are increasingly ingenious at sourcing
cheap, reliable ingredients from companies with which they
collaborate to 'co-create' processed food (Blythman, 2015).
Substitutionism uses micro-ingredients like flavourings and
colourings to impart different food qualities from 'mouthfeel
to sweetness to sometimes bombastic flavour' (Swinburn et al.,
2019, p. 128) and improve shelf-life. Flavorists or 'culinary
chemists' are playing an increasingly important role in the
plant-based meat sector, where they are intent on masking the
'vegetal green notes' in pea protein and the 'beany notes' in
soy, while strengthening the 'mineral, musky, charry, umami
flavours' associated with meat. Recreating perfect food tex-
tures is the current horizon for food researchers, as the secrets

of taste have already been discovered thanks to chemical analysis of aromas including 'phantom aromas' – traces undetectable to human awareness that trigger heightened perception of taste through embodied memories.

The high level of chemical tinkering of plant-based meat clashes with recent trends such as local and farm-to-table dining (Barber, 2014). Marion Nestle says fake meat 'fully meets the definition of ultra-processed food' (cited in Reiley, 2019a). Britain's leading investigative food journalist, Joanna Blythman (2019a), argues that the 'plant-based push gives [fast food] a get-out-of-jail-free card'. She notes that a McDonald's vegan Veggie Dippers meal contains nuggets presenting 40 ingredients – KFC's 'zero chicken vegan burger' has a similar pedigree. Perversely, substitution also facilitates the manufacture of diet, health and functional foods or 'nutraceuticals' that claim to improve on the nutritional value of the original food (Scrinis, 2013). As Julie Guthman suggests 'surely, if the same policy environment that allows the production of "fattening food" also produces non-digestible or noxious (but cheap) diet food, we have to rethink the problem statement' (Guthman, 2011, p.128).

The novel foods industry struggles to find palatable terms to describe the products in development. On one hand, they are battling traditional meat producers who are pushing for stricter marketing regulations concerning the use of words such as 'meat', 'burger', 'sausage', 'jerky' and 'hot dog'. By 2019, over 30 states in the US had enacted legislation dictating that 'any food product containing cell-cultured animal tissue or plant-based or insect-based food shall not be labelled meat or as a meat product'(Reiley, 2019b). Some purveyors of novel foods agree, albeit for different reasons. Leonardo Marcovitz, founder of FFW (Food for the World) argues that advertising plant-based protein products as 'alternative meats' perpetuates 'a "gold standard" for [meat] being the king, and [imitation] products being secondary' (Southey, 2019a). Memphis Meats' Eric Schulze said his company will continue to use the term 'cell-based meat' because it is 'factual, inclusive and neutral', and 'transparency for consumers is paramount'. But Barb Stuckey, the chief innovation officer at Silicon Valley food innovation and development firm Mattson, believes 'consumers don't want to eat science. When you talk about cells, you leave the realm of deliciousness, which makes it very difficult to sell people on an idea'. She prefers 'cultivated meat' because it is

> ... a story and a visual analogy that likens the production of cultivated meat to the propagation or cultivation of plants, where you take a cutting and put it in a warm and fertile environment, grow it and harvest it...we want to tell the story in a way that helps this amazing transformative technology see the light of day and not get stuck in the position that GMOs are today.
>
> (cited in Watson, 2019a)

The history of GMOs has lessons for cultured meat promoters. They both share consumer fears about safety and contamination, attracting the moniker 'Frankenfoods', and their industry structures bear striking similarities. These include initial feverish enthusiasm for the new technology, financial pressure for R&D to bring products to market and then scale up production, and reliance on risk framing and perception rather than 'shifts in technological, economic or agricultural realities' (Mohorčich & Reese, 2019, p. 3).

A clear 'GMO divide' has grown to distinguish between locally-based food economies and global, fossil fuel-based industrial agriculture (Krimsky, 2019, p. 132). Though originally developed by small biotech start-ups, GMOs are now synonymous with giants like Monsanto, Dow and Tyson. The business model of these transnationals drives them to realise large investments in R&D by developing lucrative patents on products to capture market share. Companies like Monsanto have been targeted for focusing on these short-term profits over environmental and social concerns (Robin, 2010). The seed economy stands out as a point of contention. Companies maintain 'technology stewardship agreements' under which farmers are granted limited licences. Growers must sign the agreement and are prohibited from sharing seeds. Monsanto must be granted access to crop land to take samples and inspect seeds. The removal, by law, of farmers' rights to own and fully control their seeds along with the patenting of those seeds remains the ultimate expression of *biohegemony*, the methods corporations have used to take control of the food system. Patents themselves are at the heart of the new colonialism (Shiva, 2007, p. 273).

Consumer resistance against GMO technologies has been higher in Europe than the US, where the industry has benefited from a softer regulatory approach and press media have been less critical. Nevertheless, pressure applied to suppliers and retailers in North American food supply chains has prevented widespread commercialisation of some products. For example, foodservice chains expressed concern that french fries sales might be adversely affected by disquiet over GM potatoes (Herring & Paarlberg, 2016). As a consequence, cash crops including soybean, canola, cottonseed oil and corn are the most modified to date. Ingredients such as soybean oil, corn starch and corn syrup derived from these crops are used in processed food rather than grains, fruits and vegetables for direct human consumption.

Advocates argue that the technologies are the solution to food insecurity and can assist in mitigation of climate change in the form of 'climate-friendly' crops that use nitrogen more efficiently and increase the potential for soil carbon storage. Opponents counter these advantages with concerns over long-term harm to ecosystems and human health. The so-called Green Revolution transformed India's Punjab with spectacular yield increases, followed by dry aquifers, salinized soils and farmer suicides; disaster brought about by a failure to reconcile technologies with local conditions (Patel, 2009). Gene editing can lead to deformations and the emergence of allergenic proteins in livestock. Heavily subsidised GM corn is found in 90% of tortillas in Mexico, 30% of which contains residues of glyphosate, the cancer-causing compound in Monsanto's RoundUp Ready herbicide (Suppan, 2017). The contamination of Mexican maize landraces by GMOs has had a huge impact not only on biodiversity but on Mexican diets and an entire food culture (Gálvez, 2018). Free trade agreements like NAFTA enable patent-holding multinationals to sue seed savers and peasant farmers, and exercise a form of soft power founded on food dependency that has disrupted not only peasant livelihoods and ecosystems but an entire society.

> Contemporary biotechnology disrupts maize at the molecular level, by inserting bacterial gene sequences into the maize genome with unpredictable consequences. Neoliberalism disrupts maize agriculture at the social and economic levels, by displacing and fragmenting rural communities, by reducing their members to the status of precarious individual labourers who have no option but to feed themselves with low-quality, potentially poisonous imported maize.

(Cota, 2016, p. 4)

While they are prone to the same, overreaching heroic narrative of 'feeding the world', most plant-based and cultured meat start-ups consider themselves open and transparent on their use of GMOs in products. Impossible Foods declares

> ... *genetic engineering is an essential part of our mission and our product. We've always embraced the responsible, constructive use of genetic engineering to solve critical environmental, health, safety and food security problems, and have long advocated for responsible use of this technology in the food system.*

(Impossible Foods, 2019)

Backed by scientific champions like Michael Eisen, Professor of Molecular and Cell Biology at UC Berkeley (and shareholder), Impossible Foods is leveraging the challenges of global warming, water shortages, land degradation and a growing population to promote their products. Eisen argues that the threats posed by a changing climate compel us to adopt

> ... *powerful new tools that allow us to modify DNA in order to generate specific valuable traits, rather than waiting for them to be delivered by the random winds of mutation.*

(Eisen, 2018)

Nevermind the 'adaptive technological rationality' of Mesoamerican and other pre-industrial societies who refined wild and domestic crops within self-organising systems over thousands of years (Cota, 2016, p. 15; see also Pascoe, 2014 & Yunkaporta, 2019).

Rather than just focusing on 'corporate-driven eco-efficiency' (Frédéric, Hite & Gregorini, 2020) perhaps those in the industry should consider the values that underpin their efforts to create the next 'novel' food. They might contest the logic of invoking technology as a *deus ex machina* to fix a food system built on overproduction, planned obsolescence and waste (Foster & Clark, 2020, p. 249). As Wendell Berry writes

> ... *epic feats of engineering require only a few brilliant technicians and a lot of money. But feeding a world of people year to year for a long time requires cultures of husbandry fitted to the nature of millions of unique small places – precisely the kind of cultures that industrialism has purposely disvalued, uprooted and destroyed.*
>
> (Berry, 2017, p. 337)

These 'cultures of husbandry' are exemplified by the complex webs of ritual and relations that resist and endure in traditional Indigenous foodways.

MORE-THAN-FOOD

Although Indigenous communities are rarely still reliant on wild food for their food security, traditional food production and consumption remain vital to their well-being, their relationships with territory and the transmission of culture, philosophy and values to future generations. While embedded through necessity in the wider political economy, Indigenous people engage in traditional food production as it 'provides a sense of uniqueness and participation in a community' and a vital sense of belonging and place. Potawatomi environmental philosopher Kyle Powys Whyte (cited in Hoover, 2017, p. 217) notes that the 'relational responsibilities' that ensure a tribal community's 'collective continuance' and 'comprehensive aims at robust living' include traditional food cultures and economies.

The Mi'kmaq, whose lands stretch across the north-eastern coast of what is now North America, possess a traditional food culture heavily focused on meat consumption. The moose is considered to be 'the chief of all land animals'. It plays a large role in coming of age rituals where hunting 'served as a young man's entry into adulthood by signalling that he possessed the skills to support a family and the patience and maturity to participate in political councils' (Robinson, 2016, p. 264). As the moose is sacred, hunters are expected to show respect through traditional ceremony. Pondering 'is the moose still my brother if we don't eat him?' Margaret Robinson explains

> … the ceremonial interaction with animal spirits is seen as a continuation of our interaction with embodied animals. To fail to show respect to an animal spirit risks future food security, since animal spirits are believed to reincarnate (increasing the animal population) and to communicate with others of their kind, reporting on how we have treated them in life and death.

(Robinson, 2016, p. 265)

Robinson is 'excited by the prospect that in vitro meat will reduce or eliminate animal slaughter and suffering' but also concerned that

> … the advent of in vitro meat might mean that moose meat comes to be seen as a thing – a despirited secular object – rather than the sacrificial gift of a brother and a friend. If the mechanisms by which sacrifice is construed – the moose permitting themselves to be caught—is removed from the equation, and moose meat is grown and harvested,

will our relationship with our food, and with the moose from which it originates, still be grounded in respect and gratitude?

(Robinson, 2016, p. 271)

Further, Robinson worries whether forgoing the wild meat might cement 'assimilation to settler values and further distance us from our traditional intimacy with animals'. The potential shift from 'a communal value system to an individualistic value system' would be damaging to 'the experience of being a community, and of engaging in communal activities that express and reinforce our identity as a people'. Accordingly, hunting and harvesting a moose is 'about more than just meat'. Technologies like in vitro meat present 'a future where I need not interact with anyone to have moose meat – especially not the moose itself' (Robinson, 2016, p. 272). Robinson suggests two possible paths forward, should the technology be adopted. The first is applying the same cultural protocols for the collection of traditional plants and medicines to cultured meat. The second means incorporating the new technology into Mi'kmaq oral traditions and stories where 'the choice to eat in vitro meat could embody our own regret at animal death, and at our failure to live out the value of Netukulimk', a key tenet of Mi'kmaq philosophy which refers to being a protector of other species (Robinson, 2016, p. 275).

This sensitive and generous openness acknowledges the wider benefits of low-input and low-emission meat and dairy production systems for ecosystem health and underlines their vital socioeconomic role in Indigenous and rural communities (Hocquette, 2016). Widespread expectations that new innovations like cultured meats and plant-based proteins are the solution to the problems of our food systems dismiss more immediate and significant sustainability gains that are already available. Wild and biodiverse cultivation set-ups contribute to soil health, sustain farmer livelihoods and feed communities successfully throughout the majority world. In many of these systems, wild game and livestock play an important role not only in supplying protein but in nurturing soil health without contributing excessively to climate change (Lengnick, 2015; Provenza, 2018; Dalrymple & Hilliard, 2020). *Nature* reports that widespread adoption of a semi-vegetarian or flexitarian diet might reduce GHGs by as much as 52%. This research cites that low-impact, sustainably produced beef, for example, can generate lower GHGs than crops like coffee or cocoa beans produced in deforested areas (Springmann et al., 2018).

Beyond simply tracking carbon emissions we need to transition to a food system that restores, rather than damages, nature and ensures a healthy, resilient food supply into the future. It can be argued that

> ... *the priority given to meat alternatives with limited sustainability potential is not just a problem of technological optimisation of production systems but also a second order problem of problem framing, network building, assumptions about innovation and economic-technological imagination.*
>
> (van der Weele, Feindt, Goot, Van Mierlo, & Boekel, 2019, p. 511)

Framing the problem of our food systems in terms of economics and geopolitics only promotes competition over collaboration. There is a dangerous edge to the race between countries, and corporations, to become global leaders in the geo-engineering of our food. For instance, RethinkX threatens that if the US fails to adopt transformative food technologies, countries such as China will capture the 'health, wealth and jobs' that accrue to those leading the way

(Tubb & Seba, 2019, p. 9). At a time when we need global cooperation, this is an alarming call to action.

Technologies to develop cultured meat and ultra-processed foods may have altruistic applications in the future. However, they further distance us from the origins of our food and leave us 'ensnared in the traps of technology under the guise of contentment' (Provenza, 2018, p. 170). Proclamations that 'the future of food will be high-tech tell us little about the values of the food system we're building for future generations' (Patel, Jayaraman, Boyd, Shute, Perls & Carpenter, 2017). Thomas Pausz highlights 'headless' material activism that 'like magic tricks, can leave us with a sense of childish enchantment' while missing 'what is really at stake'. 'What are the current and emergent technologies involved and who controls them? What is the role of myth-making and narrative design in the substitutive movement?', he asks. The point of meat substitutes is 'not to change the consumer's habits by convincing them of accepting a new philosophy based on ethics or a rational discourse'.

> *Quite the opposite, the idea is to comfort the consumer by creating the perfect substitutes, so we can carry on eating "burgers" and "chicken" without having to suffer from a change of taste or texture in the substitution…The power of these apparently superficial substitutions is that they do not have to be presented as a massive cultural change to operate… Isn't meat production a symptom of a wider malaise in society? If this symptom disappears, what happens to the roots of the problem?*

(Pausz, 2015)

We don't need new products, we need a new paradigm.

4

CHANGING OUR WATER WAYS

THE END OF HISTORY

The President of the Maldives, Mohamed Nasheed, donned his wetsuit to go to work on 17 October, 2009. At an underwater cabinet meeting, he signed a national commitment to become carbon neutral within 10 years. This stunt, engineered to address what Rob Nixon calls the 'drama deficit' of climate change, renders visible how its slow violence impacts most on those who have contributed least to global emissions (Nixon, 2013, p. 264). If the seas rise as forecast, the Maldives' entire population of over 500,000 citizens will become climate refugees. It is one of 43 island nations facing the 'end of history' if decisive action on climate change is not taken.

Islands provide a stock analogy for finitude and appear in the extinction narratives of species and civilisations throughout history and fiction. Food is always implicated in these tales of collapse (think of the dodo, think of Rapa Nui). Islands, in a sense, insert themselves between two responses to our warming condition – a retreat to localisation (fine if you are not underwater) and submission to the hyperobject. Island writers and artists have a long experience to bring to the fast and slow violence of empire; 'they have a complex history of staging paradoxical relations between the local and the global, posing allegorical antinomies or paradoxes for figuring the island as a world' (DeLoughrey, 2019, p. 10). A new vocabulary is emerging in Oceania to describe climate change or *draki veisau*. Cartographies are being redrawn. The Alliance of Small Island States (AOSIS) is leading the world in climate change mitigation.

As fisheries are the 'cornerstone' of food security for most island states, it is vital to understand the climatic impacts on stocks (Johnson, Bell, & De Young, 2013). The 22 Pacific Island Countries and Territories (PICTs) cover an area of more than 27 million square kilometres and have a combined population of 9.9 million people, anticipated to reach 15 million by 2035. More reliant on fisheries for economic development, government revenue, food security and livelihoods than anywhere else in the world, seven of the PICTs receive up to 40% of taxes from tuna fishing licences sold to distant fishing nations and another five countries/territories derive up to 25% of their GDP from industrial fisheries and fish processing. Fish consumption is often two to four times the global average and supplies 50–90% of dietary animal protein in rural areas. Across the region, 50% of households in surveyed coastal communities earn their first or second incomes from fishing or selling fish.

This reliance makes the smaller island economies highly vulnerable to climatic changes that affect fisheries and aquaculture. El Niño-Southern Oscillation (ENSO) rainfall, sea surface temperatures and ocean acidification are on the increase, preceding an anticipated rise in sea-level of 90–140 cm by 2100. Ocean currents and eddies in the region, especially those near the equator, are expected to change, impacting primarily skipjack tuna (*Katsuwonus pelamis*) which is expected to shift eastward and to higher latitudes as increasing ocean temperatures make the western equatorial Pacific unsuitable for spawning. The decline of coral reefs, seagrasses and mangroves is expected to be more severe. For example, the combination of increased sea surface temperatures, decreased solar radiation, changing rainfall patterns and possible increases in cyclone intensity could reduce seagrass area by as much as 50% by 2100. The decrease in coastal fisheries production associated with these changes is expected to be between 10–30% (Johnson et al., 2013).

The impacts of climate change will vary across the region. Increases in rainfall in the equatorial Pacific are projected to be very high, whereas decreases are projected in the southeast of the region. Tropical cyclones (projected to be less frequent but more intense), weakening of the equatorial and northeast trade winds and strengthening of southeast trade winds will drive significant changes in ocean circulation. The western Pacific warm pool will expand and the Pacific equatorial divergence zone is expected to contract, shifting the feeding grounds for tuna to the east. While the combined effects of climate change are expected to potentially enhance habitats for freshwater fish and invertebrates, this growth will be offset by declines in coral reef fish, requiring careful management of tuna fisheries to balance food security between the east and west islands. This is essential to reduce reliance on calorific imported foods that are causing some of the highest levels of NCDs in the world (Bell & Ganachaud, 2013; Charlton et al., 2016).

With its high rates of obesity and diabetes the Pacific reveals the insidious, slow violence of colonial foodways perpetuated in brutal nutrition transitions. Patterns of commerce that profit wealthy nations have replaced traditional healthy Islander diets of fish and vegetables with ultra-processed foods, sugary beverages and the offcuts of the global meat trade.

> *Our story is about a fatty, cheap meat eaten by peoples*
> *in the Pacific Islands, who are among the most*
> *overweight in the world. Lamb or mutton flaps – sheep*
> *bellies – are often 50% fat. They move from first world*
> *pastures and pens in New Zealand and Australia,*
> *where white people rarely eat them, to Third World*
> *pots and plates in the Pacific Islands, where brown*
> *people frequently eat them – and in large amounts.*
>
> (cited in Winders & Ransom, 2019, p. 77)

Relieving pressure from exhausted global fisheries is a growing challenge as seafood not only becomes a more important source of animal protein around the world but is increasingly exploited as animal feed and as an ingredient for fertiliser, thus contributing to emissions. Remarkably, seafood makes up a higher proportion of international trade than all terrestrially produced meats combined – 35% compared to 27% (FAO, 2018). Aquaculture is growing in scale to support capture fisheries which are almost fully exploited. However, like the CAFOs, it props up the unsustainable food-fuel complex by requiring grain-based fish foods, and presents socio-economic and environmental challenges for host communities, which will be explored in this chapter. Stepping out of this lock-in involves looking to lower trophic levels in the aquatic food web, and reckoning with the impacts of our appetite for fish.

GREEN, SLIMY AND NUTRITIOUS

'If I had one silver bullet for the future of sustainable food, it would be green and loaded with algae', says Mariliis Holm, co-founder and Chief Science Officer of Nonfood, an(other) early stage sustainable food start-up based in Los Angeles.

Nonfood's first product, the Nonbar, is comprised of 37% algae and 'aquatic plant' ingredients and delivers nine grams of 'clean' protein, antioxidants, vitamin A, calcium and absorbable iron (Kite-Powell, 2018). Holm believes algae meets the criteria that 'truly innovative and successful food products need to fulfil…tackle planet Earth's health, improve human health through nutrition, be affordable and above all taste delicious'. She is not the first person to be convinced that algae-based proteins are the 'the future of food' (Kateman, 2019). The first Algae Mass-Culture Symposium held at Stanford University in 1952 announced algae's arrival as a serious research topic in countries including the US, Germany and Japan. But it has been a food source for humans for millennia. Aztecs were observed by the Spanish to consume a blue-green cake made from Arthrospira, also known as spirulina (Caporgno & Mathys, 2018) and First Nations people of coastal British Columbia, including the Gitga'at and Tsimshian peoples, were so reliant on *Porphyra abbotae*, a species of red algae, that they named the harvest month of May after it (Turner, 2011).

Broadly defined as 'oxygen-producing, photosynthetic, unicellular or multicellular organisms excluding embryophyte terrestrial plants and lichens', algae include seaweed and microalgae (Bleakly & Hayes, 2017). They possess many climate-friendly characteristics – higher protein yield per unit area compared to terrestrial crops and marginal requirements for freshwater and arable land – less than 2.5 square metres per kg of protein compared to 47–64 square metres for pork, 42–52 square metres for chicken and 144–258 square metres for beef (Caporgno & Mathys, 2018).

The health benefits of algae match their environmental credentials. Spirulina is the most highly consumed microalgae species due to its high protein content and value as a treatment for hypertension and hyperglycemia. It contains 180% more calcium than an equivalent quantity of milk, 670% more protein than tofu and 5100% more iron than spinach (Bleakly & Hayes, 2017). The amino acid profiles of both Chlorella and Arthrospira are similar to conventional protein sources such as eggs and soybean, making algae a potential replacement for unsustainable soy imports. Microalgae are also a source of carbohydrates, polyunsaturated fatty acids and essential minerals. They can be incorporated into a range of existing food products including dairy, pasta, bread and biscuits to increase protein profiles and are gaining a reputation as 'innovative functional food products' in their own right (Caporgno & Mathys, 2018). While protein levels are highly variable among species and methods of extraction are yet to be scaled-up, the potential of algae as a sustainable food solution is clear. Self-proclaimed 'wellness company' iWi is growing algae on a massive scale on 900 acres in the New Mexico desert. CEO Miguel Calatayud suggests that one day algae will appear 'in every single food that you make on an everyday basis'. The *Nannochloropsis* algae that iWi is cultivating is 40% protein and can produce about seven times the amount of protein as soybeans on the same amount of land while releasing oxygen into the air. It is grown in long ponds called 'raceways' that are exposed to the sun as opposed to closed systems which require photobioreactor technology (Crane, 2018).

Ironically, while the production of algae is now global and farmed seaweed has more than doubled worldwide since 2000, the forests of the seabed have been disappearing. The impacts of industrial pollution and climate change are already evident in decreasing stocks. Underwater forests of seaweed are the foundation of polycultures of herbivorous fish and invertebrates that underpin marine foodwebs. They store carbon dioxide and convert nitrates and phosphorus from waste to plant tissue and remove heavy metals and other contaminants from the water, making seafood safe for larger species and human consumption. Their patterns of decline are widespread and alarming in speed and scale, and accelerating with nutrient run-off and the removal of predators at the top of the food web through overfishing. Restoration efforts around the globe have enhanced underwater biodiversity in degraded areas. In their inspiring project 'Operation Crayweed', Adriana Vergés and her collaborators sought to make visible, and remedy, the complete disappearance of *Phyllospora comosa*, or crayweed, from Sydney's coast in Australia (Vergés et al., 2020). In their multidisciplinary study, they note that the rates of success in restoration of seaweed forests are lower than those of land-based habitats, recognising 'this is due to the logistical challenges associated with working in open and dynamic marine systems that are at the mercy of currents and waves but is also due to the relative invisibility of what happens beneath the waves' (Vergés et al., 2020, p. 242).

To a great extent, the hidden bounty of our seas that lies in algae and seaweeds has, like the impacts of our consumption of fisheries, remained 'out of sight' and therefore 'out of mind' (Rayfuse, 2020). Our favourite seafoods tend be high trophic level specimens like salmon, tuna, cod and haddock. Filter feeders, algae and even tiny fish – all low on the trophic scale – are not only economically viable but more sustainable and nutritious than their predators (Bonhommeau et al., 2013; Duarte et al., 2009; Duarte, Marbá, & Holmer, 2007; Olsen, 2015). Trophic levels expose how the *food chain* concept is misleading. Ecosystems are not a linear series of simple steps, unlike our fragile global food supply chains. The notion of the *food web* demonstrates how many organisms feed on the prey of different trophic levels in the ocean – and the large extent to which we depend on marine species to sustain terrestrial food systems (FAO, 2019). Elspeth Probyn notes 'we all eat fish, albeit in circuitous ways' (Probyn, 2016, p. 5). Twenty-five percent of the global fish catch is reconstituted into products ranging from fertiliser to animal feed, including food for other fish. In its complexity, the marine ecosystem resists simplification. Oceanic depths are particularly difficult to conceptualise, resisting 'flat terrestrial maps that position humans as disengaged spectators'. To consider the ocean, we must submerge ourselves in 'entanglements of knowledge, science, economics and power' (Alaimo, 2017, p. 107). A good place to start is the supermarket.

THE LOSS OF ABUNDANCE

The modern supermarket has its very own 'ecology' based on a fine balance between supply and demand that does not suit the sea. While terrestrial proteins like beef can be produced through calculable and addressable risks, watery ecosystems are much less manageable. Even less so in a changing climate where rising temperatures and fluctuations in prey fish populations can set off trophic cascades. Paul Greenberg suggests that two incompatible systems run side-by-side – 'the human-focused, need-driven system where demand remains constant; and the diverse, disparate natural marine system that varies from year to year as a result of a plethora of uncontrollable variables' (Greenberg, 2010, p. 169). Emblematic of our efforts to find a source of fish 'so abundant that massive, consistent deduction will not cause an imploding of the stock' is the collapse of *Gadus morhua* or cod. Privatisation, monopolisation and industrialisation manifested in the emergence of corporate factory ships which outnumbered, outpaced and outfished their small, local competitors. With advanced trawling technology including immense bottom-dragging nets with 'rock hopper' attachments that enabled the penetration of sheltering banks where the fish mated and spawned, the commercial fishers ate themselves out of business. The New England fishing grounds, Georges Bank, were closed in 1994, leading Mark Kurlansky to lament 'is this the last of wild food?' (Kurlansky, 1998). The loss of cod became a global metaphor for the loss of abundance because it was part of diets of so many ordinary people; even a marker of identity and integration in countries such as England, where fish and chips became the national dish (Panayi, 2014). Its white flaky texture and taste became synonymous with the Filet-O-Fish sandwich in 1962, available to you for a cost of 25 cents. Based on 2010 consumption rates, humanity 'needs' 40 billion pounds of cod-like fish annually – the size of the entire Georges Bank codfish population – *every year* (Greenberg, 2010, p. 142).

Upon the collapse of cod (defined as when 90% or more of the population is gone), a replacement was obviously required. The world's largest seafood buyer Unilever, already under fire from the ocean conservation movement, was under pressure to find a 'sustainable' alterative. In a stroke of marketing brilliance, it embarked on a partnership with the World Wildlife Fund (WWF) and jointly established the Marine Stewardship Council (MSC), devoted to setting global standards for sustainability in fish stocks (Greenberg, 2010, p. 170). The hoki or whiptail, a New Zealand fish with a cod-like consistency and taste, was identified as a suitable target by Unilever. Hoki is a deepwater fish which has risen commercially as stocks of Orange Roughy, a former favourite, have declined. The latter, with a lifespan of more than 100 years and slow reproduction rates, could not keep pace with demand. Hoki mature more quickly, with a lifespan of 25 years, and are therefore viewed as more sustainable. In 2001, hoki was granted MSC certification, a process contested and subjected to an appeal by groups including the New Zealand Royal Forest and Bird Protection Society who described it as 'a farce' (cited in Greenberg, 2010, p. 171). By 2009, New Zealand had slashed the allowable catch of hoki from 275,000 tons to 100,000 tons after the US-based watchdog Blue Ocean Institute gave the fish an unfavourable orange rating (on a scale from green to red) in terms of sustainability, noting considerable bycatch including seabirds and fur seals. The Institute described the fishery's management as 'driven by short-term gains at the expense of long-term rewards' (Broad, 2009). Despite these warnings, New Zealand's West Coast hoki fishery had collapsed by 2018 (Young, 2018).

This case, and that of other species such as the Alaskan pollock, illustrates how despite a myriad of certification standards and promotional campaigns to protect fish there is a profound 'disconnect between production and appetite, knowledge, cultural practice and the political will embedded in them and embodied in consumers' (Sharp, 2020, p. 128). In the global supermarket ecology, a constant supply of fish is essential, regardless of natural limits, and the dominant firms exercise their power to influence regulators and watchdogs. When cod prices rose in response to scarcity McDonald's supply chain leader on fish, Gary Johnson, said 'I had no idea what I specifically needed, but I knew the end result was that we had to get a sustainable assured supply of fish without harming the environment' (Langert, 2019, p. 121). Teaming up with Conservation International (CI), collaborators on the epic failure that was the first environmentally themed Happy Meal 'Discover the Rainforest', Johnson set about engaging with suppliers. He hired a blokey marine biologist, Jim Cannon, who 'won over the tough fishermen…he bellied up to the bar like your very best mate' and got the suppliers to agree to sign up to a fish scorecard following a basic stop-light methodology (Langert, 2019, p. 123). According to Bob Langert, a former McDonald's executive, this was 'the power and influence of McDonald's supply chain working at its best, balancing cost, quality, value, and, now, this fourth element of sustainability' (Langert, 2019,

p. 124–125). In 2013, the company signed up to the MSC standards, stating:

> *McDonald's collaboration with the Marine Stewardship Council is a critical part of our company's journey to advance positive environmental and economic practices in our supply chain. We're extremely proud of the fact that this decision ensures our customers will continue to enjoy the same great taste and high quality of our fish with the additional assurance that the fish they are buying can be traced back to a fishery that meets MSC's strict sustainability standard.*

(Langert, 2019, p. 125)

The story of cod and hoki traces a narrative arc that fisheries have historically followed – 'rich harvests, overfishing, and collapse' (Green, 2020). Since Hugo Grotius flagged the possible need to limit fisheries back in 1608, we have received repeated warnings (Rayfuse, 2020). The numbers baffle. A 2003 study estimated that we have fished 90% of large fish from the oceans. In 2013, one bluefin tuna sold for US$1.7 million dollars (Servigne & Stevens, 2020, p. 52). The fisheries sector is considered 'crucial in meeting FAO's goal of a world without hunger and malnutrition' (FAO, 2018, p. vii), and it is pulling its weight. Consumption of seafood per capita has skyrocketed from less than three kilograms in the 1950s to approximately 20 kilograms in 2015 (Mills, 2018, p. 1273). Largely due to 'relatively stable capture fisheries production, reduced wastage and continued aquaculture growth', global fish production reached 171 million tonnes in 2016, 88% of which was consumed by humans. Total production was valued at US$362 billion, with US$232 from aquaculture (FAO, 2018, p. vii). In 2015, fish supplied 17% of animal protein consumed and provided 3.2 billion people with almost 20% of their animal protein (FAO, 2018).

The likelihood of obtaining more 'biomass' from the oceans in 'a way that does not deprive future generations of their benefits' (European Commission, 2017) seems low. Stocks fished at biologically unsustainable levels increased from 10% in 1974 to 33.1% in 2015, mainly in the Mediterranean and Black Seas, Southeast Pacific and Southwest Atlantic. Up to 59.7% of fish stocks are 'maximally sustainably' fished, and only 7% of fish stocks are under-fished (FAO, 2018, p. 6).

The key to functioning ecosystems is maintaining species at specific levels of abundance; to recycle nutrients through a biome requires strength in numbers (Giggs, 2020). Stress on stocks has profound and potentially devastating impacts on marine ecosystems that are being compounded by ocean acidification, 'climate change's evil twin' (cited in Alaimo, 2017). As larger fish disappear, the phytoplankton they eat flourish. They bloom briefly, and upon death are consumed by bacteria who consume oxygen, causing anoxia in the depths of the ocean. If the anoxic layer rises to the surface, hydrogen sulphide, a gas destructive of the ozone layer, might make air unbreathable. These stagnant 'Canfield Oceans' would destroy life on Earth as they did in the Proterozoic era (Canfield, 1998; Servigne & Stevens, 2020). This is an extreme example of trophic cascading, whereby 'one shock causes other shocks down through the complex webs of life in the sea' (Probyn, 2016, p. 247). The ocean is now 30% more acidic than at the dawn of the industrial era and we need to refer back 55 million years to find a comparable measure. According to Christopher Sabine, leading researcher in ocean acidification, oceans absorb 22 million tons of CO_2, the equivalent of approximately 22 million Volkswagen Bug cars, per day. He notes the rate of absorption is still rising, and 43% of the overall amount has been added in just the past 20 years (cited in Mueller, 2017, p. 213). This is the source of ocean acidification which kills pteropods or sea butterflies, tiny snails that, through their abundance, provide the foundations of the entire oceanic food web. Acidification dissolves the shells of the pteropods, which contribute to the diets of krill, whales, mackerel, carnivorous zooplankton and many species in between.

Forms of 'invisible environmental harm' (Alaimo, 2017, pp. 112–113) such as the destruction of these tiny marine snails are a form of slow violence being played out in vital marine ecosystems such as coral reefs. According to the Australian Government Great Barrier Reef Marine Park Authority (GBRMPA, 2019, p. 1)

> ...only the strongest and fastest possible actions to decrease global greenhouse gas emissions will reduce the risks and limit the impacts of climate change on the Reef. Further impacts can be minimised by limiting global temperature increase to the maximum extent possible and fast-tracking actions to build Reef resilience.

At 1.5°C coral reefs, already experiencing bleaching and ocean acidification, are forecast to decline a further 70–90%. An increased frequency of bleaching events will prevent reef recovery. If GHG emissions continue at their current rate, annual bleaching events are anticipated by 2044 (GBRMPA, 2019). While altering the productivity, structure and composition of reefs and other ecosystems on which fish depend for food and shelter climate change is having profound impacts on fish physiology, behaviour, growth, development, reproductive capacity, mortality and distribution; it also enables competitive species, such as the Pacific oyster, and their pathogens, to spread to new areas which can lead to mass mortalities of many aquatic species (Brander, 2007).

Climate change is the catalyst for a sequence of interlinked disasters that are clear signs of 'an ocean ecology in crisis', according to Lesley Green (2020, p. 14). She numbers among these episodes the Red Tide event of 1989 in False Bay, on the Cape coast in South Africa, when a hypoxic swamp of dead abalone, crayfish, mussels and sea urchins washed up on the shoreline; a cascade of bizarre mass strandings and changing diets and migration patterns among species of lobsters; outflows of raw sewage and storm water exacerbated by extreme weather events including droughts; and scientific studies revealing potentially endocrine-disrupting and carcinogenic levels of household cleaning products and pharmaceuticals in filter feeders like limpet and starfish. This is not only an ecological crisis but a crisis of governance brought about by 'a capitalist accounting of the sea' (Green, 2020, p. 23) that has driven its expansion into fragile liminal zones and onto land in the form of aquaculture.

THE BLUE REVOLUTION

Like its terrestrial counterpart the Green Revolution, the Blue Revolution heralds the application of technology to the taming of wild species; in this case the abundant resources of oceans, rivers, lakes, Arctic regions and wetlands. Under the guise of taking pressure off wild marine ecosystems, aquaculture has expanded into a massive global industry. In the 1980s, the World Bank and Asian Development Bank invested US$200 million a year in aquaculture projects like shrimp ponds that replaced mangrove forests in the Philippines, Thailand and Ecuador. Carp and tilapia farms sprawled over the flood plains of the Ganges, the Irrawaddy and the Mekong

Rivers. Aquaculture output doubled from 1975 to 1985 (Boychuk, 1992). Today, an acre of Vietnamese *Pangasius*, or tra, will generate half a million pounds of fish, as opposed to 10,000 of cod. Tra's propensity for rapid reproduction and its high stocking capacity has made it the fourth most common aquaculture product in the world (Greenberg, 2010). Like tilapia, which originated in the Nile but is now farmed widely in Latin America, tra thrive on a diet of corn and soy which also improves their flavour; on an organic waste and algal diet their flavour resembles mud. Their corn- and soy-based diet props up unsustainable grain production in other parts of the food web. Other less intended consequences include the spreading of these invasive species across fragile ecosystems like river deltas. Of great concern to human health is the heavy reliance of fish farms on antibiotics to combat the infectious diseases that thrive in all types of intensive feeding operations. The combined impact of these drugs in aquaculture and land-bound contaminated waterways is the emergence and spread of antimicrobial resistant (AMR) bacteria which are respon-sible for over 35,000 deaths in the US, 33,000 in Europe and 58,000 in India annually. These numbers are higher in South East Asia and are bound to increase with warmer temperatures (Reverter et al., 2020).

The diseases and parasites that the antibiotics are designed to manage have profound impacts on species in the wider ecosystem. Wild salmon in the Pacific Northwest, for example, have been infected by sea lice when they swim near infested caged salmon (Bailey & Tran, 2019). Salmon is a keystone species in aquatic and terrestrial environments, meaning it plays a central role in supporting ecosystems. *Salmo salar*, 'the leaper' in Latin, inspired a sense of awe in cave painters that persists today. In 1653, it was crowned the 'king of freshwater fish' (Mueller, 2017). The decline of wild salmon is attributed to overfishing and ocean acidification but also logging, river damming, grazing, pollution and urban sprawl. In Washington, Idaho, Oregon and California, salmon now live in only 40% of their historic habitat. Before the Gold Rush of the mid-1800s, salmon spawning habitat along the Sierra Nevada could have stretched across the continent and back again; by the 1960s, 97% of it was gone, sacrificed to hydroelectric dams and irrigation for agriculture (Reisner, 1986). Now almonds compete with salmon in drought-stricken California. In 2015, the Columbia River suffered 'unprecedented' salmon losses due to a reduced snow pack and higher river temperatures, leading Teresa Scott of the Washington Department of Fish and Wildlife to announce it as a 'wake-up call and a dress rehearsal for what fishery managers years from now will be dealing with on a regular basis' (cited in Mueller, 2017, p. 269).

Today, Norway dominates the global salmon industry – in 2010 it was worth 31.5 billion kroner and supplied 97 countries. By 2014, export revenues reached 46 billion kroner, triggering the suggestion of professor in fishery economics Rögnvaldur Hannesson that Norway's wild salmon should be sacrificed for their 'domesticated cousins':

> *We should perhaps ask ourselves what we want wild salmon for? If wild salmon get in the way of the fish farming industry, then I say we must be ready to sacrifice wild salmon. The industry creates great values and jobs along the entire coast. It is an important business branch, one that is important to keep. We need not feel pity for the upper class that will miss a playroom; surely they'll find some corresponding amusement.*

(cited in Mueller, 2017, p. xvii)

While this says more about Professor Hannesson than it does about salmon farming, it also nails the philosophy of an industrial food system committed to the absolute commodification of nature. Martin Lee Mueller describes this as a 'semantics of separation' that enables the objectification of a species that, equipped with a 'large and diverse genetic base, survived every upheaval of the two northern oceans for millions of years' (Mueller, 2017, p. 49). In farms, salmon are reduced to biomass with no 'intrinsic value'; they become a 'concrete equation' where 'living beings equal flesh; flesh equals mass; mass equals numbers; numbers equal economic performance' (Mueller, 2017, p. 33). They are a 'resource', a 'product', a 'brand', terms that 'consistently and systematically exclude other ways of knowing' (Mueller, 2017, p. 53). Like chickens and livestock, salmon have been bred for rapid growth, larger body size, higher stress tolerance and disease resistance. Escapees breed with free-ranging mates, reducing the survival rates of their hybrid offspring. These have greatly reduced adaptation to climate change. As in their land-bound counterparts, the CAFOs, the fish farms demonstrate how

> ...*modern industrial practices obstruct possibilities to encounter other animals in their individuality, to form robust social bonds with them, or to develop social practices that would strengthen such bonds further. They create a perceptual lifeworld in which feelings of empathy, compassion, even love for another being, are rendered redundant.*
>
> (Mueller, 2017, p. 51; see also Pachirat, 2013)

We are further removed from the salmon by GM technologies that splice the antifreeze hormones of ocean pout, a cold-water eel, with a Chinook salmon growth hormone gene and transplant this into an Atlantic salmon. This produces a fish that looks like a salmon but is actually a hyperactive, perpetually feeding growth-machine that can reach market size in half the time of its real-life rivals. Patented by AquaBounty of Waltham, Massachusetts, as the AquAdvantage (trademark) salmon, the fish is defined as 'identical in every measurement and every respect' on the basis of 'substantial equivalence' — US Food and Drug Administration (FDA) speak for new foods that 'show the same composition and characteristics of the ordinary food'(cited in Mueller, 2017, p. 31). Suitably vague, this paradox is an example of what Vandana Shiva (2016) describes as the 'ontological schizophrenia' of an industry that profits by branding dubious technologies as new and novel but also natural and safe.

As production, and innovation, increases and more farmed salmon appears on our supermarket shelves, it might yet replace tuna as the 'chicken of the sea'. The fish farm, like the feedlot and the gene lab, is an example of 'the omnipresence and sophistication of technological spaces' which perpetuates the fantasy of the Anthropocene 'that human ingenuity really is the measure of all things' (Mueller, 2017, p. 224). Even more so, as fish farms appear tranquil; we cannot see what lies beneath. This 'disconnect between knowing and seeing works in the industry's favour' (Mueller, 2017, p. 54). Further, empathy is hard to come by; for the land-bound, it is harder to establish what Berglijot Børresen calls an *ichthyocentric* or fish-eye view. On the contrary, says Mueller.

> Given the salmon's very long sojourn inside this ancient planet, it would seem to be more parsimonious and precise to think of salmon not as objects, not as commodities, but as elders.
>
> (Mueller, 2017, p. 274)

This expression of our human responsibility to fish, and fish pluralities, is mirrored in Zoe Todd's reflection on Indigenous human-fish relations in Canada that 'fish carry stories in their bones' and that without them we are left not just physically but *philosophically* hungry' (Todd, 2018, p. 72).

The negative impacts of fish farming occur out of the narrow purview of the average seafood eater. Replicating the logics of factory farming and industrial meat processing plants, fish are caught and bred to 'satisfy whimsical gustatory predilections rather than the requirements of sound ecologically based husbandry' (Greenberg, 2010, p. 13). Aquaculture answers our apparent urge to 'eliminate all wildness from the sea and replace it with some kind of human controlled system' rather than going to the trouble of ensuring wild environments are 'understood and managed well enough to keep humanity and the marine world in balance' (Greenberg, 2010, p. 12). Currently the world's fastest growing food sector, aquaculture is subject to the same dynamics of vertical integration and concentration as the meat and dairy industries, with six of the world's top 12 seafood corporations expanding investment into feed mills, processing and harvesting small pelagic fish for fishmeal and fish oil (FAO, 2018). Cargill's investment in Norwegian salmon is strategically plotted to expand their position in feed supply, the single largest cost in aquaculture production.

The benefits of the Blue Revolution, like those of the Green, are unequally distributed. Consumers who can afford to purchase fish have an abundance of choice and high quality. Rural people displaced by aquaculture industries are left with fewer options. Even those who do find employment in these facilities are likely to experience food insecurity as they cannot access the market (Bailey & Tran, 2019).

THE PLIGHT OF THE FISHERS

If fish get no respect, think a while about the small-scale fisher. Small-scale fisheries are defined as fisheries 'where fishers operating from the shore or small fishing vessels use simple methods to catch fish from inland or coastal waters' (World-Fish, 2017). More than half a billion people worldwide depend on fisheries as 'economic and social engine[s], providing food and nutrition security, employment and other multiplier effects to local economies' (FAO, 2015).

Unsurprisingly small-fishers face a number of 'overlapping exclusions' including competition from large-scale mechanised fishing operations, increasing pressure to serve global markets and ocean-grabbing (Mills, 2018, p. 3). *The Global Ocean Grab* report (TNI, 2015b) defines ocean-grabbing as 'the capturing of control by powerful economic actors of crucial decision-making…including the power to decide how and for what purposes marine resources are used, conserved and managed'. The two leading global fishers' movements, World Forum of Fish Harvesters and Fish Workers and the World Forum of Fisher Peoples (2015), supplement this definition to include 'the exclusion of small-scale fishers from access to fisheries and other natural resources'. On the seas, property rights are incomplete, access to the resource is 'open' and fish are 'fugitive resources' of which it is difficult to claim ownership (Hannesson, 2008). Efforts to protect fisheries and mitigate the impacts of climate change including 'no fishing' zones have joined large-scale commercial fishing fleets as the biggest challenges to small-scale fisheries (Mann, 2020a). Increasingly problematic are 'blue carbon' initiatives, whereby governments and corporations buy credits by investing in the protection of coastal areas in order to offset their emissions. These can depoliticise the problem of inequitable distribution of access and resources, and tend to

> *...concentrate power and generate wealth for a*
> *relative few, fail to recognise the diverse rights and*
> *flexibility of small-scale fishers, and interrupt*
> *mechanisms that previously distributed economic*
> *and food security benefits broadly – particularly to*
> *those most in need.*
>
> (WorldFish, 2017; see also Taylor, 2015)

The disabling and discriminatory elements of these governance mechanisms are contested by an emerging 'fisheries justice movement' engaged in a collective struggle for 'inclusion, equal rights, and the democratisation of access, ownership, and control of natural resources and fishing territories' (Mills, 2018, p. 3). Describing conservation measures such as profit-driven eco-tourism and marine protected areas as 'ocean-grabbing under the cloak of "sustainability"' the food sovereignty movement aims 'to link up the struggles resisting against land, water, ocean and green grabbing – all of which indeed intersect'. La Vía Campesina insists that the 'true guardians of the fisheries resources' – First Nations and small-scale fishers – should be given back control of their waterways (La Vía Campesina, 2017, p. 3).

At the heart of this struggle is the *aquaecological* approach to fishing. Based on species-specific equipment and techniques; following life cycle and breeding patterns; protecting coastal ecosystems; adhering closely to catch limitations; and participatory governance, aquaecology shares with the *agroecological* approach to farming a strong commitment to empowerment of local communities and a resistance to corporate control (Mann, 2019). This reflects the root of both concepts in food sovereignty as 'the right of people to healthy and culturally appropriate food produced through ecologically sound and sustainable methods, and their right to define their own food and agriculture systems' (Declaration of Nyéléni, 2007). As a politically motivating concept, aquaecology resists the simplification or dumbing down of ocean politics by interrogating who fishes where, how, and how much. This is central to a geopolitical context where 'overfishing has become framed in a cruel scenario that sees commercial fishers painted as the rapists of the sea, and lauds a so-called ethics of refusing to eat fish' (Probyn, 2020, p. 28). Elspeth Probyn interrogates one case which links the flight of Senegalese migrants across the Mediterranean to the illegal fishing activities of Chinese fleets in the exclusive economic zones (EEZs) off the West African coast. The Africa Progress Panel's estimation of losses to Senegal alone amounts to US$300 million, or 2% of the country's GDP. Local communities literally starve as the tiny sardinella on which not only the Senegalese but most Western Africans rely for food security are processed into fishmeal for the aquaculture industry. This scenario illustrates how global governing frameworks such as the United Nations Convention on the Law of the Seas can 'set in motion a spatialisation of the seas that places coastal nations of the Global South at risk of losing their precious more-than-human marine resources' (Probyn, 2020, p. 38). Further, they support the global trade regime as the spoils of the Chinese fleet are distributed globally – 6% of all fish products consumed in the EU in 2010 came from China. Meanwhile 'the Chinese fisherman and the Senegalese migrants are united by the lack of fish, the lack of their livelihoods and ways of life' (Probyn, 2020, p. 29).

WATER POLITICS

The battle over the world's waterways is the nexus of food security and climate writ large. As William Hague said, 'We cannot have food, water, or energy security without climate security' (cited in Brown, 2011, p. 15). Societal collapse is directly related to food shortages, many of which have been triggered by declining water supplies. In the ancient civilisation of Sumer, rising salt concentrations in the soil exposed a flaw in irrigation design, leading to a decline in wheat yields. A shift to salt-tolerant barley was not enough to stave off the collapse of that civilisation. The Mayans deforested their land to such an extent it affected regional climate. The reduction in rainfall, in conjunction with soil erosion, led to food shortages and ultimate collapse. Drought, famine and disease are inexorably linked (McMichael, 2017).

Olivier De Schutter, the former UN Special Rapporteur for the Right to Food, stated nearly 10 years ago that 'without rapid action to claw back waters from unsustainable practices, fisheries will no longer be able to play a critical role in securing the right to food of millions', noting that 'with agricultural systems under increasing pressure, many people are now looking to rivers, lakes and oceans to provide an increasing share of our dietary protein' (Schutter, 2012). Inland fisheries are in danger of disappearing, and aquaculture will experience losses of production and infrastructure due to extreme events such as floods, increased risks of diseases, parasites and harmful algal blooms. In 60 years, Lake Chad has decreased in area by 90%, from 26,000 square kilometres in 1963 to less than 1,500 in 2018 (UN, 2018). Changes in the availability and trade of fish products have important geopolitical and economic consequences for the tropics, but also the South Pacific, Pakistan, Iraq, Morocco and Spain.

Water-based 'food bubbles' artificially inflate the price of grain while draining aquifers. When these burst, we see food riots and the emergence of failed states. Exploitation of global water supplies is leading to critical shortages in some of the world's flashpoints. These 'hydrological basket cases', as Lester Brown (2011) calls them, exist in some of the most populous and volatile regions of the world including Syria,

Yemen and Pakistan. Further, climate stress drives poverty and desperation that feeds insurgency and interstate rivalry in hot spots ranging from Afghanistan to Mexico to Zimbabwe. Those societies most dependent on agriculture and fishing, many of which lie between the Tropics of Capricorn and Cancer, represent what Christian Parenti calls 'the tropic of chaos' (Parenti, 2011). These 'economically and politically battered post-colonial states' are the vanguard of a changing climate. They are also vital to the world's food security. Their trajectories illustrate that 'the climate crisis is not a *technical* problem, nor even an *economic* problem: it is, fundamentally, a *political* problem' (Parenti, 2011, p. 226).

Pacific politics provide a case in point. In November 2018, our Prime Minister Scott Morrison announced an AU$2 billion infrastructure financing plan as part of Australia's 'Pacific step-up' to counter rising Chinese influence in the region (Martin, 2019). In August 2019 he was in Tuvalu at the Pacific Island Forum, opposing sections of the Tuvalu Agreement on emissions reductions, coal use and climate funding on the grounds that he is 'accountable to the Australian people' (cited in Clarke, 2019). Tuvalu's former Prime Minister Enele Sopoaga acknowledged Australia's financial support but insisted on the need to reduce emissions above all else: 'No matter how much money you put on the table, it doesn't give you the excuse not to do the right thing. Cutting down your emissions, including not opening your coal mines, that is the thing we want to see' (Zhou & Walsh, 2020). Professor Rosemary Lyster, Co-Director of the Australian Centre for Climate and Environmental Law, agrees. She says

> ...*there can't be a genuine 'step-up' in the Pacific until the government says out loud: 'Coal does hurt us – Australia and everyone in the Pacific and*

everyone around the world'. To say otherwise is to showcase a 'pathological, ideological opposition' to the reality that the burning of coal is a major driver of global climate change.

(Lyster, 2019, p. 18)

The pathology of climate denial is the biggest threat facing our Pacific neighbours.

5

RECOVERING FOOD WISDOM

SILENT SICKNESS

The average American body presents traces of 29 pesticides. Herbicides like glyphosate are ubiquitous and up to 11% of organic crops in the United States are cross-contaminated. 'Tolerable intakes' and rates of exposure to these poisons are based on 'outdated science' (Provenza, 2018, p. 35). The circulation of toxins throughout our ecosystems occur apace through the grasshopper effect, where persistent organic pollutants (POPs) like DDT and carbofuran 'hop' through a series of evaporation and deposition events, accumulating in the landscape and wild food sources, and in our bodies (Hird & Zahara, 2017, p. 129). Globally, but disproportionately in the Global South, farmers and food workers are subjected to poisons, without personal protective equipment (PPE), causing chronic illness, cancers and birth defects. Half a century since Rachel Carson observed 'in a very real and frightening sense, pollution of the ground-water is pollution of water everywhere' (Carson, 1962, p. 53), the toxicity in our food system is widely acknowledged by scientists and policy-makers yet insufficient action is being taken to rein it in. Use of synthetic fertilizer rose by a factor of 10 between 1950 and 1998, while pesticide use climbed forty-fold from the mid 1940s to the 1970s (Clapp, 2016, p. 53).

The production of cheap food and its proponents has always required the 'suppression of political dissent' (Patel & Moore, 2018, p. 151; see also Sinclair, 1906; Hightower, 1975). Observe how Carson was dismissed variously as a barren and hysterical female, a luddite and a communist, following the publication of *Silent Spring* (1962). Explaining how pesticides contaminate soil and water and their link to human and ecosystem health, Carson presented the futility of just focussing on the curing of diseases like cancer, rather than eliminating their causes. The greatest threat she posed to the growing agribusiness sector was her ability to share her understanding of deep ecology with a wide audience. She was a champion for sophisticated biological controls that could target pests in a way that chemical poisons could not, comparing the 'chemical barrage...hurled against the fabric of life' to a crude 'cave man's club' (cited in Souder, 2012, pp. 353–354).

The true cost of polluting factory farms and the vast, poisoned monocultures required to produce raw ingredients for the foods that are slowly killing us 'never appears on the menu' (Schlosser, 2002, p. 9). Meanwhile the pursuit of 'wellness' is a multimillion dollar industry populated by health-fixes like Goop, SoulCycle and Moon Juice – the very existence of which broadcast that 'the contemporary world is toxic, and that to endure or thrive within it requires extraordinary measures of self-regulation and self-purification' (Wallace-Wells, 2019, p. 188). As we continue to degrade our natural environments and ignore the impacts of climate change, the 'purity arena' is likely to expand, providing solace through conscious consumption and self-improvement. This is no substitute for political action. Not only it is time 'to reacquaint ourselves with the defining characteristics of food that keeps us, and the planet, healthy' (Blythman, 2020a); it is time to join the resistance against a corrupt food system. This means disengaging from a 'mono-cultural mindset' that displaces ecologically sound and truly sustainable food production and nutritional wisdoms based on generations of traditional learning (Shiva, 1993).

RELEARNING RESILIENCE

A socio-ecological understanding of resilience concerns the capacity of a system to withstand disturbances; the 'coping mechanisms and adaptive capacities that provide the means to overcome the exposures and sensitivities associated with vulnerability' (Doherty, Ensor, Heron, & Prado, 2019, p. 4). Adaptive capacity is built through the decentralisation and sharing of diverse resources and knowledges just as biodiversity builds ecosystem resilience. The loss of on-farm biodiversity is directly linked to agricultural intensification and the consolidation of production into fewer large farms. Seventy-five per cent of the world's crop diversity was lost between 1900 and 2000, most since the inception of the Green Revolution (Clapp, 2016, p. 52). Supply chain vulnerabilities are pronounced in the streamlined industrial production system whereby farmers don't produce their own seed, relying on corporations to provide seeds that boast high-germination rates and antifungal coatings as protection against disease. Modern seed supply chains, dominated by just three companies, are highly susceptible to biosecurity breaches and genetic vulnerabilities (Raasch, 2017). The vertical integration of these global supply chains has not only created a 'treadmill of production of more goods and services' but also 'facilitated the creation of systems that stimulate and drive demand with little, if any, consideration of the lasting environmental costs' (Friel, 2019, p. 65).

The race to maximise agricultural outputs has led to the maximising of agricultural inputs, putting farmers in a double-bind. From the early days of the family farm, a source of 'cheap, docile, and flexible labour' (Vivero-Pol, 2017, p. 126) farmers have carried the risks of what is a risky enterprise. Now food processors and retailers exert downward pressure on farmers whose only means to profit is by increasing outputs. They are frequently misrepresented as collectively responsible for the GHG emissions produced by agriculture. But extending responsibility for global warming to all farmers is the corollary of blaming humanity as a whole for global warming as not all farming practices produce excessive GHG emissions; '*agricultural inputs* produce greenhouse gas emissions' (NFU, 2019, p. 6).

Global warming is revealing just how vulnerable the crops and livestock we have thus far relied on are to changes in temperature and rainfall, declining soil health, pests and pathogens. Regions of monocultures like the US Corn Belt provide superhighways for pathogens and future warming will compromise any remaining natural defence systems (Philpott, 2020). Food safety is compromised through complex international supply chains making outbreaks of contamination hard to trace (Bové & Dufour, 2001; Grandin, 2020). Pressure on ecosystems, and especially the interaction of humans or livestock with wildlife, through overexpansion, deforestation and urban sprawl is triggering the emergence of novel viruses like COVID-19 (UNEP, 2020a; 2020b). Livestock serve as an epidemiological bridge between wildlife and human infections, while factory farms are breeding grounds for a range of deadly pathogens, including influenza variants (Wallace, 2016); in the war of words over the origins of COVID-19 it is important to note that 'agribusiness barns and coops are just as prone to infection as wet markets' (Salcedo Fidalgo, 2020, p. 18).

The expansion of chicken production highlights many of these threats. Globally we consume 65 billion chickens per year. Intensively raised, they produce relatively low amounts of emissions and waste per unit of meat, but negative externalities including high antibiotic use, ammonia pollution and deforestation through soy production for feed make them unsustainable. Transporting chickens internationally increases the chances of spreading zoonotic viruses, those that cross species barriers like the avian influenza A (H5N1), to new countries (Shaffer, Lindsay, Araneta, Raman, & Fowler, 2017). The Global South serves as 'a waste receptacle for the developed world' with off-cuts dumped in markets as a cheap product for consumers, affecting the livelihoods of local poultry farmers (cited in Freshour, 2019, p. 123). The modern chicken provides fewer nutrients, more fat and less Omega-3 because of genetic selection for breast meat, reliance on grain-based feed and lack of access to natural forage (Salazar, Billing, & Breen, 2020). It is not part of a healthy diet or a just food system. To disrupt the chicken economy,

> ...is to change the meat economy of the planet and everything it affects: land use, water use, waste disposal, resource consumption, the role of labor, concepts of animals' rights, and the diets of billions of people.

> (McKenna, 2017, p. 308)

This is the scale of change required to recover local, sustainable and resilient food systems or, as Philip Ackerman-Leist (2013) puts it, 'rebuild our foodsheds'.

NUTRITIONAL WISDOM

Knowledge and practices of food-based medicine have been lost in the distancing of eaters from the origins of their foods. Colonists disregarded the wisdom possessed by Indigenous peoples, shamans, healers and herbalists who practice therapeutic plant medicine. Pharmocopoeias that date back centuries can be found across regions and cultures including native North American, Australian Aboriginal, Indian Ayurveda and traditional Chinese societies. The loss of social norms and rituals around food gathering, cooking to increase palatability, digestibility and reduce toxicity, and preferences for different foods uniquely suited to specific landscapes has led to food sensitivities that conflict with our biological time-clocks. We are 'out of place' when we eat globalised diets.

The solution is to reconnect with what Fred Provenza calls 'the wisdom of the body' (Provenza, 2018, p. 84). An expert on livestock intelligence who spent much of his career following and recording the foraging habits of rangeland animals, Provenza invokes the notion of 'nutritional wisdom' as a pathway to better health. This knowledge is necessary to contest a food industry that combines synthetic flavours with fats and refined carbohydrates to stimulate our appetites while reducing the nutritional variety in our foods. As discussed in Chapter 3, different artificial flavours give the illusion of variety. We crave fat and sugar when our bodies actually need an array of phytochemicals that our ancestors evolved eating. Today, only 15 plant species are relied upon for 90% of global food consumption (Provenza, 2018, p. 36). Big Food exploits our 'propensity to generalise from past experiences' to train us to prefer artificially flavoured foods, including those presented as healthy options. This occurs when we link learnt flavour–feedback relationships from high-calorie foods and beverages to similarly flavoured versions that contain fewer calories, for example, artificially sweetened drinks. Science writer Mark Schatzker calls this process of stimulating appetite and fooling the body through 'obesity-inducing food intoxicants' that mimic our nutritional requirements the Dorito Effect (Schatzker, 2015). This leads to metabolic imbalances and the noncommunicable diseases (NCDs) now endemic to modern life.

The way food is produced in industrial agriculture, for maximum yield and generic appearance, strips it of nutritional value. The application of off-farm sources of nitrogen, phosphorus and potassium sacrifices phytochemical richness, which has declined 5% to 40% in 43 fruits, vegetables and grains across four decades. Early picking and shipping prevent natural ripening on the vine, further depleting both flavour and nutritional content (Provenza, 2018, p. 31). Broccoli might be purchased 7–10 days after harvest, a time in which 75% of health-promoting phytochemicals, 50% of vitamin C and most natural sugars and antioxidants have depleted. Further, many modern cultivars have fewer of the phytochemicals that create flavour compared with heirloom varieties – concentrations in tomatoes, for example, might vary a thousandfold from conventional species (p. 33). Foods made blander through the speeding up of production include meat, dairy, and factory-farmed poultry, now twice the body weight of birds 50 years ago (p. 130).

Deficiencies in industrially grown fruits and vegetables can be traced directly to depleted soils, where biodiversity originates. Herbicides, pesticides and fungicides kill many soil organisms that share complex symbiotic relationships. Insects, fungi and bacteria breakdown waste and catalyse nutrients; they also dissolve rocks into valuable minerals including iron, calcium and zinc which are absorbed by plants. Termites play the same role as worms, aerating soil in marginal dry environments. In applying deadly chemicals, we lose not only individual species but also the 'complex microbial worlds that make life possible' (McFall-Ngai, 2017, p. 51). Our natural and most economical carbon sink is soil, where mycorrhizal fungi and microbial interactions create soil and sequester carbon. Soil carbon restoration could store up to one billion tons of atmospheric carbon per year, offsetting 8–10% of total annual carbon dioxide emissions and one third of new carbon that would otherwise be released (Schwartz, 2013, p. 5). 'Carbon trading is something that has been going on for millennia in our soils. It underpins the health of our whole ecosystem' says Australian soil scientist Christine Jones (cited in Schwartz, 2013, p. 30). After 20 years trying to promote and educate the industry as to the biological dynamics of soil carbon sequestration, Jones realised that the science wasn't the problem; it was

> ...the implications of the enormous industry that depends on us not finding solutions to problems in agriculture...[problems] that generate income for the ever-expanding ancillary industries including the manufacture of synthetic fertilizers, herbicides, insecticides and fungicides.
>
> (cited in Schwartz, 2013, p. 40)

Broad-acre monoculture landscapes and their chemical inputs have profound ecological implications for conservation of biodiversity and the sustainability of agroecosystems (Saunders, 2016, p. 94). Intensive agricultural practices including heavy harvesting and tillage impede ground-nesting species; fertilizers, herbicides and intensive grazing reduce flora; and insecticides are lethal to a range of creatures outside their target species, including birds. Agrochemical solutions offer inadequate replacements for traditional farming practices such as crop rotation that encourage soil fertility and reduce pathogens. Declines in the abundance and species richness of solitary bees, for example, have been linked to the consistent planting of cereals over years. Pollinators supply a quarter of our calories through their industry. They are indirectly responsible for meat production as pollinators of alfalfa, grains and seeds. Coffee yields would drop 50% without bees (Newman, 2019). In the United States alone, they contribute US$19 billion to crop production annually, triggering President Obama to set in motion a national strategy including research and habitat restoration. The mystery of CCD has led to a search for high-tech solutions to a bee-less future – the engineering of new virus-resistant super-bees to accompany the human and mechanised pollination systems that already proliferate in some regions of the world including China. But if 'we put our faith in a high-tech fix, we are ignoring the bees' environmental wake-up call' (Benjamin & McCallum, 2009, p. 12). The protection of diverse and abundant food and nesting resources through 'the preservation of semi-natural elements and the development of less intensive farming practices' (Le Feon et al., 2010, p. 100) is essential not only to the survival of native species but also human health (Grauer, 2010).

Different regional realities and the varying health needs of livestock and eaters require a diversity of production systems and even more so in a changing climate. Multifunctional and agroecological approaches, promoting diversification at the field, farm and landscape scales, increase resilience. Livestock play an important role in these diverse ecosystems where 'soil-building, grass-loving graziers' are vital 'ecological service providers' (Greenwood, 2018). Here, animals are more than commodities to be consumed but essential components of an ethical food chain. They are '"living assets", the fundamental sources of food, nutrition, livelihoods, jobs, incomes, savings

and much more' (Ramsden, 2019, p. 6; see also Dalrymple &
Hilliard, 2020; Lengnick, 2015). Their role in food security
cannot be overestimated. Grazing livestock are the most
effective way to convert grass into protein for up to one billion
people globally – 70% of the rural poor are dependent on them.
In some countries, the livestock sector accounts for up to 80%
of GDP (Rust, 2019). Organic meat and dairy farmers suggest
that rather than eating plant-based products made from
industrially grown, ultra-processed soy, pea, mung bean and
rice proteins, we should be encouraging ecologically sound
rotational systems, permanent pasture and conservation
grazing. Actively avoiding the use of anti-worming agents and
antibiotics, these farmers cultivate soil health so manure feeds
earthworms, bacteria, fungi and invertebrates such as dung
beetles who play a crucial role in aerating, rotating, fertilising,
hydrating and detoxifying the soil. In response to critics of
livestock's methane emissions, organic farmers claim these are
lower in biodiverse pasture systems (Tree, 2018).

Unfortunately, the impacts of climate change on extensive
grazing ecosystems, such as those described above, will make
intensive systems 'the more favoured choice' at a time when
rising incomes, hyper-urbanisation and cultural preferences
are driving demand for animal products (Rust, 2019). Yet
forms of regenerative agriculture, holistic management and
permaculture as methods of 'inculcating healthy, living soils'
offer an antidote to the ills of both our food systems and our
health (Massy, 2017, p. 5). Farmer and author of *Call of the
Reed Warbler: A New Agriculture, a New Earth,* Charles
Massy, describes these methods as 'ecologically and socially
enhancing'. He says that by adopting the 'organic mind' over
the rational or 'mechanical' mind we are making a choice to
work with nature rather than overpowering it (Massy, 2017,
p. 27). This might sound utopian until we weigh the costs of

denuded soils and a warming climate on our own health. As Dr Charles Northern said in 1936: 'It is simpler to cure sick soils than sick people, which shall we choose?' (cited in Schwartz, 2013, p. 95).

FROM TRANSITION TO TRANSFORMATION

Challenging a food system built on colonial exploitation, dispossession and fossil fuel philosophies represents a significant cultural shift. It means disrupting what Shiva calls *eco-imperialism*, a construct under which 'corporations gain increasing control of the Earth's resources – energy, water, air, land, and biodiversity – to continue to run the industrialized globalised economy' (Shiva, 2008, p. 15). This resistance is captured by food sovereignty, defined in the Declaration of Nyéléni (2007) and cited in Chapter 4. Advancing common-based alternatives that incorporate socio-biodiversity, food sovereignty is much more that an alternative food system; 'while it produces food', it also fosters '(agro)/biodiversity, polycultures, closed metabolic cycles and ecosystem restoration, coupled with labour and decision-making based on communality, reciprocity, consensus, equity, and intersectional social justice' (Figueroa-Helland, Thomas, & Pérez Aguilera, 2018, p. 183). A rights-based approach emphasising access to resources and the right to define local agricultural systems and markets it 'pushes back against capitalism' by insisting that food is treated 'not as a commodity but as a common good' (Vivero-Pol, 2017, p. 17). Food sovereignty contests forms of political and economic blackmail whereby, for example, small farmers and Indigenous landowners are granted subsidies to facilitate their participation in monocultural industries like forestry. By stepping out of this conventional production system and creating a 'self-controlled and self-managed resource base', small-scale farmers engage in a form of 'co-production of man and living nature that interacts with the market' (van der Ploeg, 2008, p. 23). This supports a model of agriculture that follows a 'healthy logic of reproduction of social and ecological relations, as opposed to the degrading and disabling force of capitalist agriculture's dynamic of under-reproduction of social labour and ecosystems' (McMichael, 2013, p. 131). It contests the impacts of predatory food and agriculture markets that present farmers with a 'Faustian bargain of notoriously incorrigible inequalities' (Akram-Lodhi, 2013, p. 106). These now showcase technologies for climate-proofing agriculture – a new 'profit frontier' for agroindustry in the form of 'climate ready' genetically modified, patented seeds (McMichael, 2009; Taylor, 2015, p. 106).

Food sovereignty 'displaces a West-rest dichotomy', accommodating diverse perspectives in different locales and engendering 'do-it-yourself' strategies and tactics ranging from direct action to legal demands and lobbying activity (Dunford, 2017, p. 155). It is a serious contender against a dominant system that 'shuts down alternative worlds by enclosing land', and demonstrates how 'intercultural exchanges have given rise to a common demand for a world in which the diverse agricultural systems through which people produce food in their own territory are possible'. Advocates of food sovereignty assert the right to food is indivisible from other human rights including gender and racial equality, housing, health and education. The urgency of this project has escalated dramatically with COVID-19. For La Vía Campesina member *Unión de Trabajadores Agricolas Fronterizos*, based on the Mexican-US border in El Paso, Texas, the pandemic has highlighted the vulnerability of migrant workers without health benefits. They also warn of the political opportunity the virus provides governments to control populations through restricting movement, closing borders and strengthening anti-migration policies (Unión de Trabajadores Agricolas Fronterizos, 2020).

In a bizarre twist for these same governments, COVID-19 has revealed small-scale, local food systems as a strategic counterpoint to corporate food empires and their precarious global supply chains. The self-organisational capacities of peasant markets are emerging globally, for example, in Porto Alegre, Brazil; the French Basque Country; and *Campi aperti* (Open Fields) in Bologna, Italy. In these cases farmers and their communities are managing socially distanced markets and home deliveries for those in lockdown (van der Ploeg, 2020). As such they are becoming primary circuits in food provisioning, and are central to regional economic, and environmental, recovery.

Agroecology, a form of land stewardship which co-produces with nature and cultivates resilience, is central to food sovereignty and essential to addressing the challenges posed by global warming. Communities practicing agroecology in countries including Zambia and Malawi avoided the worst impacts of the food price crisis of 2008. Empirical research has revealed that agroecological methods assist in sustainable recovery after hurricanes and floods by promoting vegetation that protects topsoil, retains moisture and limits erosion, reducing economic losses in comparison to those experienced by industrial farms (Holt-Giménez, 2002). A radical, large scale shift to agroecology is needed to limit the rise in temperature to less than two degrees Celsius by 2100 (Watts, 2020). This action was taken by Cuba, in the Special Period following the collapse of the Soviet bloc (1991–2000). Forced to contend with a fossil-free future, Cubans implemented a process of *socio-economic adaptation* to new conditions. What was key to their success was an urban community structured around the *barrio* where multigenerational households relied upon each other. Urban agriculture became a 'self-help local movement' that took over vacant and abandoned plots of land (Friedrichs, 2013). The Farmer to Farmer Agroecology Movement (MACAC) united farmers in an energy transition project across the entire country. Its diffusion was facilitated by the well-established National Association of Agricultural Producers (ANAP) promoting social processes of learning through *diálogo de sabres* or 'dialogues of knowing' based on values of collectivism, solidarity and cooperation (Martínez-Torres & Rosset, 2014; Méndez, Bacon, & Cohen, 2013). Puerto Rico is currently applying the *campesino-a-campesino* (farmer to farmer) methodology to develop 'solidarity bridges' between producers, health workers and the wider community and reclaim the archipelago's food sovereignty (Tramel, 2020, p. 30).

The Canadian National Farmers Union (NFU) is playing a less radical but leading role in addressing the role of agriculture in global warming. *Tackling the Farm Crisis and the Climate Crisis: A Transformative Strategy for Canadian Farms and Food Systems* (2019) provides a blueprint for low-input agriculture that will not only lower emissions but will potentially save the national farming sector. Canadian farm debt has doubled since 2000, 'a clear message that the 40+ year experiment in high-output, high-input, high-cost food production has been a bust for farmers' (NFU, 2019, p. 11). Farmers receive 5% of their earnings – the rest go to the agribusiness corporations that supply fertilizers, machinery, fuel and credit. Off-farm work and tax-payer funded support programs are essential to staying on the farm. Canada has lost two-thirds of its young farmers (those under 35) since 1991. The NFU's project to minimise petro-industrial farm inputs by scaling up regenerative agriculture and on-farm generation of renewable energy demonstrates a commitment to a new production paradigm. These farmers are leading by example and lobbying for government-led transformation of their national food system. Among the measures and policies they recommend are rejecting governmental proposals to extend exports; diversifying production systems to support organic, holistic and agroecological methods; reverting to natural sources of fertility; shifting from fossil fuels to renewables on-farm; reducing waste in pre-and post-harvesting; minimising transportation; setting aside land for resting and alternative use; and creating an administrative body to support farmers in the transition to regenerative, Minimum-Input No-Till (MINT) agriculture. The report's compelling message is that to remain viable 'we must reimagine, restructure, rewire, and retool our farms and food systems' (NFU, 2019, p. 15).

While food sovereignty was initially formulated by a peasant movement demanding state-led land reform, it has broadened conceptually to include Indigenous struggles over territory (Rosset, 2013; Mann, 2014). This is not uncomplicated, as rights-based approaches can undermine cultural distinctions and reinforce the state's control over Indigenous people. They certainly don't restore Indigenous homelands in countries including Australia and Canada. Indigenous sovereignty is 'inherent and collective' and 'infused with interconnected autonomy nurtured through relationships with land' (Kammal, Linklater, Thompson, Dipple, & Ithinto Mechisowin Committee, 2015, p. 565). It is impossible to separate rights to land and cultural integrity in addressing Indigenous collective well-being. Decolonising activities like reclaiming land for hunting, fishing, growing, gathering food and the education of youth in local food programs are vital to reclaiming not just food sovereignty but sovereignty over resources, land and culture for groups such as the *O-Pipon-Na-Piwin Cree* Nation (OPCN) in northern Manitoba, Canada, where the Churchill River Diversion flooded many homelands in the 1970s. An example of continued colonisation under the cloaks of modernisation and national prosperity, this case highlights the meaning of reclamation as more than 'collaboration, partnership, or infrastructural development' but also the 'stopping of practices that encroach upon the sovereignty of territories' (Menser, 2014, p. 70) including the dumping of toxins, the siting of polluting industries, excessive draw-down from water tables and the damming of life-giving waterways. The OPCN use the concept of *wechihituwin* to describe any 'means of livelihood that is shared and used to help another person, family or the community' (Kammal et al., 2015, pp. 555–556). Defined as such food and other productive resources comprise a set of relationships rather than objects or commodities, and sovereignty is not about control over but the re-establishment of relationships with land, water and wildlife.

Intercultural exchanges between non-Indigenous and Indigenous peoples' movements in Latin America have led to the incorporation of cosmovisions that emphasise that living well, or *buen vivir,* is only possible collectively through the 'interrelation of beings, knowledges, logics and rationalities of thought, action, existence and living' across shared territory (Walsh, 2010, p. 18). As such, food sovereignty is 'rooted in Indigenous revitalization and/or agroecology' (Figueroa-Helland et al., 2018, p. 183). Aligned with movements of Indigenous peoples worldwide food sovereignty contests the dispossession of and removal of people from their traditional lands and the obliteration of their foodways through being forcibly 'locked within extractive schemes that dispossess them, either through privatisation of lands or through conservation that produces expulsive migrations' (Gómez-Barris, 2017, p. 87). In a 'racialized discourse of property' not only corporates and governments but also well-intentioned NGOs contribute to the aims of 'unpeopled landscapes' which not only supports the construction of the myth of *terra nullius* but also overlooks the fact that Indigenous peoples contribute to, rather than erode, biodiversity.

HEALING COUNTRY

Country in Aboriginal English is both a common and proper noun. Country is addressed as a person; it

> ...*knows, hears, feels, smells, takes notice, takes care, is sorry or happy...Country is a living entity with a yesterday, today and tomorrow...a consciousness, a will toward life...nourishment for mind, body and spirit.*

(Rose, 1996, p. 7)

Indigenous cultures' deep ecological knowledges and connections to Country and other species have nourished them physically and spiritually for thousands of years. Customary norms and practices, notions of custodianship and stewardship obligations protect ecosystem health and manage species distribution (Williams & Hardison, 2013). For example, the Hawaiians practice a stewardship model designed to protect their islands from depletion: higher altitudes were not harvested and complex rules governed who could eat what (Newman, 2019). Totems, described as *borrungur* by the Noongar peoples of Southwestern Australia, are plants and species – and even elements like wind and lightning – adopted by Indigenous people in a process whereby 'they become you and you become them' in a deep kinship that engenders care and protection (Honeybone & McAllister, 2019; Yunkaporta, 2019). When wild killer whales in the Salish Sea off British Columbia were facing declining prey and pollution, the

Lummi people fed them Chinook salmon, stating 'those are our relations under the waves' (Giggs, 2020, p. 99).

First Nations peoples' relationships with Country share a more powerful sense of entitlement than ownership through purchase or appropriation (Goodall, 2019, p. 171). They recognise 'bonds' of responsibility to the land that are passed from generation to generation. These bonds are honoured in the 'rematriation' of seeds and foods back into communities. Rematriation describes

> ...*an instance where land, air, water, animals, plants, ideas and ways of doing things and living are purposefully returned to their original natural context – their mother, the great Female Holy Wild.*

(Prechel cited in Rowen, 2018)

The rematriation projects of groups like the Indigenous Seed Keepers Network (ISKN) of North America are about not only saving heritage seeds but mark the 'beginning of cultural sanity and healing' by keeping cultures alive, 'in the soil and our daily lives' through sharing, planting and harvesting.

It is now internationally recognised that 'agricultural practices that include Indigenous and local knowledge can contribute to overcoming the combined challenges of climate change, food security, biodiversity conservation, and combating desertification and land degradation' (IPCC, 2019, p. 29). Australia is one of the countries with the greatest potential to reduce diet-related GHGs if it seriously embraces its native foods. Aboriginal peoples' careful stewardship of the Australian continent is an example of sophisticated land management, an advanced 'proto-agriculture' that included cultivation of edible food species, food storing and harvesting, and the strategic selection and organisation of land according to its suitability for specific uses (Gammage, 2012; Pascoe, 2014). Upstream of Bourke on the Barwon river at Brewarrina are rock fish-traps that are not only among the first examples of aquaculture but possibly the oldest human-made structures in the world (Simons, 2020a).

The mosaic burning regimes, or fire-stick farming, described in Bruce Pascoe's brilliant *Dark Emu Black Seeds: Agriculture or Accident?* (2014) are now being reintroduced in Aboriginal-led carbon farming projects as pathways to intergenerational learning, connection to Country and wealth generation for communities throughout Australia (AbCF, 2020). Pascoe notes that in countering global warming, 'new ideas and methods will arise out of the very oldest land use practices' (Pascoe, 2014, p. 148). Among these are experimenting with grazing Indigenous animals and growing Indigenous crops. Aboriginal people knew that by very deliberately and systematically caring for the land, they influenced what food was available and how much. Native grains and tubers have been domesticated by Aboriginal people and proven

> ...*good to eat, good for us after 100,000 years of testing, and good for the country. Most are perennial, so there is less tilling involved, so less carbon release, plus we don't have to devote scarce water resources and fertilisers. And we reduce our dependence on using weedicides and pesticides.*

(Pascoe, 2018)

By growing Aboriginal plants and grains like warrigal greens, cumbungi (bulrush) and murnong (yam daisy),

Australia 'could reach our carbon emission targets easily'. So, Pascoe asks,

> *Why don't we? Are we that paranoid about*
> *Aboriginal claims to the land that we can't*
> *acknowledge the plants tested and domesticated over*
> *a longer period than anywhere else on earth?*
>
> (Pascoe, 2018)

He notes that where there is substantial interest in commercialising traditional foods Indigenous people must have ownership 'so large food companies don't put a brand on it and dispossess us once again...we don't want to be dispossessed twice' (Pascoe cited in Vernon, 2019).

Given colonial histories of exploitation, lack of recognition and respect for values and rights, and appropriation of resources Indigenous people are understandably wary of 'partnership arrangements' and particularly those that do not incorporate a 'spirit of mutualism and reciprocal accommodation' (Williams & Hardison, 2013, p. 542). Western understandings of traditional ecological knowledge (TEK) are frequently framed in a material, transactional sense that does not fit the spiritual beliefs and kinship relations traditional knowledge-holders share with their environments (Little Bear, 2000). For example, traditional water-related knowledge has enabled Indigenous Australians to thrive in an arid continent for over 60,000 years, yet the future of the Murray-Darling, our most vital inland water system, is of deep concern. Pascoe is particularly critical of water management characterised by corporate deal-making that favours extractive industries, including fracking, over water for communities.

> *Capitalism provides a platform for decisions among*
> *fellow capitalists but shudders under the load of*

*persuading communities over vast areas of the
country. If that were not so we would not have
reached such impasse...we would never consider
leaving a state in our Federation without drinking
water, we would not have laws which allow coal
seam gas miners to ruin a farmer's land and threaten
the very groundwater of the continent.*

(Pascoe, 2014, p. 155)

The Murray-Darling river system is a tragedy of epic pro-
portions for the millions who rely on it and particularly for
Indigenous people for whom it represents a 'cultural highway,
a massive Appian way' (De Pieri, 2019). It embraces four
states including Queensland, where a large inland geological
basin, the Galilee, spans 247,000 square kilometres encompassing 12 Indigenous language groups. Relying on the
Northern section of Australia's Great Artesian Basin, the
hydrology of which is extremely complex, the Galilee includes
the headwaters of seven major rivers. It is also the site of the
hugely controversial Adani coal mine. Central to the concerns
of the traditional owners of the mine site, the Wangan and
Jagalingou people, is the impact of the mine on the local
Doongmabulla Springs, fed by 60 tributaries and covering
more than 10 hectares. This unique ecosystem is one of the
few permanent sources of water in the area and vital to many
species found nowhere else on the planet. For a region in
almost perpetual drought, this is an essential resource for
farms and communities. Yet the state government exempted
Adani from new restrictions regulating water use in 2017. It
takes 250 litres of water to produce a tonne of coal. Only
36% of over 300 spring complexes surveyed in the Galilee
Basin at the turn of last century are still active after 120 years
of drawdown for agriculture (Bradley, 2019).

Agriculture accounts for approximately 70% of water consumption in Australia, with cotton and irrigated pastures historically accounting for the majority. Water markets are highly localised, with trade occurring between users within a single river catchment with the exception of the southern Murray-Darling where water trading between systems and across state boundaries is enabled through hydrological connectivity (ABARES, 2019). Water prices have reached their peak in the southern Murray-Darling Basin, leading to an inquiry into the water market. Foreign investors, corporate speculation, poor water-sharing policy and the introduction of new almond crops have been variously blamed for water shortages, exacerbated by reduced winter rainfall and higher temperatures directly related to climate change. In January 2019, temperatures exceeded 35 degrees Celsius over 14 days, driving demand for irrigation from 4.5 to 7 gigalitres a day (Hughes, 2019).

One of the most visible and shocking impacts of the Murray-Darling's ill health has been a series of mass death events among native fish populations. In early 2020, 20 different locations in the Macquarie, Namoi, Gwydir, Border, Barwon-Darling, Lachlan, Upper Murray, Murrumbidgee and Lower Darling rivers experienced blue algae breakouts and bushfire ash runoff (Readfearn, 2020). These episodes follow a catastrophic event across December 2018 – January 2019 when over a million fish died on a 40 kilometre stretch of the Darling River, downstream from Menindee Lakes, a vital waterbird habitat and important cultural site for the local Barkandji people. Their voices are conspicuously absent in water management. Under the NSW Water Management Act, traditional owners possess Native Title, or Basic Landholder Rights. Despite the recognition of Native Title rights for 128,000 square kilometres of Barkandji land in 2015 after an 18 year legal battle, the Barwon-Darling Water Sharing Plan provides zero allocation for Native Title (Thompson, 2019).

This is just one example of government failure to recognise and accommodate Aboriginal rights in Australia's mismanaged water governance regime. There is plentiful evidence that Indigenous-driven co-governance, distributed appropriately through families and communities, offers the best prospects for successful integration of TEK and Western science in managing social-ecological systems (Hill et al., 2012). Yet engagements in Australia are characterised by 'top-down consultation meetings which are then ignored or…"fake consultation"' (Simons, 2020a, p. 97). The consequence is 'a river system run to the edge of its ability to survive', jeopardising the future of a country that relies on it for food production (Simons, 2020a, p. 68). As an investigation into water theft, bureaucratic incompetence and political corruption proceeds, journalist Margaret Simons reflects that

> …*blaming individuals, individual communities, or the growers of certain crops obscures the larger failure – of our politics when faced with a complex challenge. Underlying that is the failure of our ability, as Australians, to recognise common interests.*

The lessons of the Murray-Darling and other ecosystems exhausted through greed and rivalry demonstrate a detachment from natural environments and the multispecies that inhabit them. This is 'an ominous sign for their survival and ours' considering that 'our decisions about nature – whether or not to nurture the landscapes, watersheds, and airscapes that sustain human communities – are based upon our relationship with nature' (Provenza, 2018, p. 107). Our reliance on these natural systems is what we have in common in spite of an 'economy invested in the narrative of competitive individualism' (Goodall, 2019, p. 7).

REALIZING *COMUNALIDAD*

In Latin America, brutal forms of extractive globalisation have spurred a resistance grounded in *pensamiento autonomía* or autonomous thought – a determination and conviction that another world is possible (Escobar, 2017, p. 16). Coupled with *comunalidad*, a non-translatable concept that captures our entanglement and interdependence with not just each other but the natural world, *autonomía* is a powerful source of energy, inspiration and creativity for local communities. Food and water are central to this counter-narrative which captured global attention with the rise of Movimiento Zapatista in Chiapas, Mexico, against the impacts of NAFTA in 1994 and struggles in defence of water against corporate ownership in Cochabamba, Bolivia, in 2001. These movements embody what Mexican sociologist Raquel Gutiérrez Aguilar refers to *entramados comunitarios* or 'communitarian entanglements'. They make visible 'diverse and immensely varied collective human configurations' that experience 'gigantic and global confrontation' with coalitions of transnational corporations and other economic powers that 'saturate the global space with their police and armed bands, their allegedly "expert" discourses and images, and their rigidly hierarchical rules and institutions' (cited in Escobar, 2017, p. 178).

Communal forms of being-knowing-doing like *comunalidad* share with the notion of the commons an alternative vision to privatisation, competition and commodification. The paradigm of food as a commons offers a way to value food and administer its production and allocation differently; it can

> *...unlock our imagination, encouraging us to design other types of policies and legal frameworks for the food system that have been so far disallowed because*

they were not aligned to the dominant narratives of capitalism.

(Vivero-Pol et al., 2019, p. 2)

In this way, commons-thinking encourages us to 'read for difference' (Gibson-Graham, 1996) and embrace the reality that 'there are no complete knowledges' (De Sousa Santos, Nunes, & Meneses, 2007, p. xlvii). It promotes 'a food politics that achieves different socio-environmental justice outcomes to those of conventional food systems' (Harris, 2009, p. 60). It recognises food as central to our diverse cultures, and insists governments not only ensure the right to food of every citizen is respected but also provide opportunities for civic engagement in and celebration of food (Vivero-Pol, 2017).

As we are all too aware, governments, institutions and political leaders often fail in times of disaster. At these times, civil society steps up,

...not only in an emotional demonstration of altruism and mutual aid but also in practical mustering of creativity and resources to meet the challenges. Only this dispersed force of countless people making countless decisions is adequate to a major crisis.

(Solnit, 2009, p. 305)

We witness this in the resurgence of commons-based cooperatives, owned and run by their members. A revival of interest in the cooperative movement was spurred by evidence of their resilience following the GFC. In France, 128 firms facing closure were successfully converted into workers' cooperatives in 2010–2011, ensuring their survival. Cooperatives supply two-thirds of rural households in India and 40% in Africa with their consumables (Hilary, 2013). In the United States, cooperatives emerged 'out of necessity' during the civil rights movement when Black Americans had to create their own services. Today, outfits like the Federation of Southern Cooperatives (FSC), an association of 75 Black and mostly farmer co-ops that share transport, marketing, storage, packing and distribution facilities, are supporting Americans through the pandemic (Nargi, 2020). In addition to grocery stores and food sharing, they embrace services including healthcare, housing, daycare and credit unions.

The motivations for starting food cooperatives include concern for community health and welfare, and the stimulation of economic activity directed towards 'social ends rather than simply the pursuit of profit – in other words towards social production' (Hilary, 2013, p. 156). Lana Dee Povitz writes that racial pride also drove the establishment of one of the most enduring New York examples, Park Slope Food Coop, which opened in February 1973. She notes how cooperatives, community-based food programs and nonprofit social service agencies provide access points for food activism as 'a powerful vector for social inclusivity, providing a sense of community as an alternative to alienation or abandonment'. They create

> ...opportunities for movement ideas to find fuller
> expression and for people to engage one another,
> collaborating, mentoring, arguing, bonding, grieving,
> and finding their bearings, often coming together
> across lines of social difference.
>
> (Povitz, 2019, p. 10; see also Kammal et al., 2015)

Recognition of local communities as vital spaces for the emergence of social leadership is essential to fostering forms of food citizenship that challenge the limited, market-based role of the consumer. After all, 'there is no way to effectively challenge placeless power without powerful places' (Taylor, 2019, p. 274). Reclaiming our agency generates energy for change from below, where we live and build solidarity; the Barcelona en Comú (Barcelona in Common) party quotes, 'democracy begins where you live' (cited in Broad, 2016, p. 268). Notions of the 'right to the city' and 'spatial justice' underpin movements towards decentralised, non-hierarchical governance where the concept of citizenship is expanded to incorporate those oppressed by capitalist urban development (McGuirk, 2014; Venturini, Değirmenci, & Morales, 2019). Connecting urbanisation, marginalisation, and surplus production and use, the right to the city is a 'protest call' towards a more equitable society (Venturini, 2019, p. 87).

The rise of this novel style of 'municipalism' is seen in the breaking away of over 350 European cities to meet the 2016 Paris Accord on emissions reduction (Taylor, 2019) and a host of complementary translocal initiatives such as the Milan Urban Food Policy Pact (MUFPP, 2015). The MUFPP recognises that cities host more than half the world's population and that they must play a major role in actions towards mitigating climate change while managing vast public infrastructure. At time of writing 210 global cities had signed the MUFPP, in doing so committing to reducing food waste, promoting healthy diets and respecting the environment, human rights and workers' dignity. By relating food access and availability to wider sets of public goods such as jobs, housing and transport MUFPP is creating new solidarities. Inviting citizen input on local issues, MUFPP's 'food cities' embrace the ethos of the participatory commons and the anarchist proposition that no one person owns ideas, knowledge or solutions. This thinking is also embedded in the methodology of food hackathons that encourage visioning of alternative futures on topics ranging from food waste management to the role of robotics and artificial intelligence in small-scale urban agriculture (Future Food Institute, 2014). In Barking and Dagenham, London, the 'Every One. Every Day.' (2020) project extends the participatory culture approach to citizen leadership of 'common denominator' activities such as fixing, trading, recycling, batch cooking and growing food. These low/no cost and commitment activities are disruptive as they have the capacity to transform food and food experiences in ways that make them accessible to all, therefore contributing to social inclusion and the development of social capital (Mann, 2020b; 2020c). They offer a vision of citizenship 'that is not based on a patriarchal birthright of exclusive privilege, not an ethnic identity or legal status, but instead is based on residency and active participation' (Taylor, 2019, p. 269). They challenge neo-liberal capitalism's 'predatory formations' (Sassen, 2014), the structures of rule that create multiple exclusions in not just our food systems but across society, focussing instead on the creation of 'new, non-exploitative forms of life' (Escobar, 2017, p. 7).

When the lived experience and local knowledge of marginalised communities becomes the basis of a co-creative design practice, design becomes 'a formidable political and material resource' (Escobar, 2017, p. 59). Radical participatory methods foster social innovation in the reconceptualisation of urban life, from the future dwellings we will inhabit to the embedding of digital technologies within human- and place-centred design. Working with subaltern groups on the frontlines of global warming, hunger and poverty in ways that strengthen their collective autonomy, transition activists address adaptation and resilience through 'the creation of grounded, situated, and pervasive design capacity by communities themselves who are bound together through culture and a common will to survive' (Escobar, 2017, p. 40). These are the 'seedbeds for transformations of capitalism' (Harvey, 2015, p. 163) towards an alternative future that captures the essence of *comunalidad* as 'a contemporary form of life that incorporates what arrives from afar, yet without allowing it to destroy or dissolve what is one's own (*lo propio*)' (Gurrero cited in Escobar, 2017, p. 177).

We discuss how we might join them in the final chapter.

6

RESILIENCE THROUGH RESISTANCE

THE SAFE AND JUST SPACE

It is ironic that the qualities most valued in the industrial food system, including Just-In-Time (JIT) inventory management that enables retailers to 'pile it high and sell it cheap', have been found wanting in the COVID-19 crisis (Pollan, 2020). A more tragic irony is that 'underlying conditions' including obesity and chronic diseases such as hypertension and Type 2 diabetes are products of our diets. The Centers for Disease Control and Prevention (CDC) reported in April 2020 that 49% of those hospitalised for COVID-19 already suffered from hypertension, 48% were obese and 28% had diabetes (Garg et al., 2020). A disproportionate burden of these illnesses is borne by Black, Latinx and Indigenous communities. This is a clear indicator of 'the classism and racism embedded within our food system' (Alkon, Bowen, Kato, & Young, 2020, p. 536). Disparities are emerging around the globe in new socio-economic indicators including capacity to social distance. Agricultural workers, such the Latinx and Haitian immigrants of Immokalee, Florida – America's tomato capital – are experiencing the highest documented infection rates, globally (Holmes, 2020; Sankey, 2020).

Weaknesses in food systems make visible social structures that contribute to inequities in health outcomes include the distribution of wealth, agency and resources, the qualities of environments, employment conditions and social exclusion. In March 2019 WHO Chief Tedros Adhanom Ghebreyesus announced the formation of the Healthier Populations division which brings together issues previously siloed across different WHO sections in a 'coordinated, intersectional approach' (Coopes, 2019). He cited the Ottawa Charter (WHO, 1986) preconditions for health ('peace, shelter, education, food, income, stable ecosystem, sustainable resources, social justice and equity') and announced

> ...globally, we need a radical reorientation of our health systems towards promoting health and preventing diseases, not simply managing them in hospitals. To do that, we need to address the reasons people get sick and die, in the air we breathe, the food we eat, the water we drink and the conditions in which we live, work and play.
>
> (WHO, 2019)

In response to the announcement adjunct Associate Professor of the Institute of Public Health at National Yang Ming University Shu-Ti Chiou tweeted, 'now social determinants and environmental determinants are in the "structure" of WHO, next we need to have political determinants to combat commercial determinants to make a true difference' (Chiou cited in Coopes, 2019).

While the original social determinants emphasised a whole of government approach and the need for civil society, local communities, business and international agencies to be involved in the effort to attain health equity as 'an ethical imperative and a matter of social justice' (Friel, 2019, p. xxvii) little progress has been made in limiting the reach of Big Food into global markets; even less so drawing the lines between these companies and their contribution to climate change. The Rio Political Declaration on Social Determinants of Health (WHO, 2011) effectively set aside threats to health posed by environmental degradation and the underlying causes of these threats (Friel, 2019). The United Nations Development Programme's (UNDP) *Sustainability and Equity: A Better Future for All* (UNDP, 2011) was an advance in terms of recognition of 'failures' to address 'grave environmental risks and deepening inequalities', highlighting connections between climate change vulnerabilities and other threats. It emphasises the lack of political power held by 'the world's most disadvantaged people [who] suffer the most from environmental degradation' (UNDP, 2011, p. 5).

In WEIRD societies, human rights to healthy environments, and the embedding of principles of social equity, public participation, and official accountability in policy, are virtually unquestioned. With good reason – there is ample evidence that supportive social infrastructure and the reduction of inequities across all dimensions of life improves collective well-being (Goodall, 2019; Klinenberg, 2018; Wilkinson & Pickett, 2009).

> *The convenient truth is that greater equality offers not only the possibility of a reduction in consumerism and status competition but also the development of a more cohesive, sociable and sustainable society.*
>
> (Wilkinson, Pickett, & De Vogli, 2010, p. 1140)

Climate-focused policies to reduce carbon emissions such as transitions to cleaner energy production and more efficient waste management also have a positive impact on collective well-being. Mainstreaming these is the challenge.

Wholesale, sectoral transformation is recommended by Kate Raworth, whose celebrated Doughnut creates a 'safe and just space' between 'a social foundation of well-being and an ecological ceiling of planetary pressure' (Raworth, 2018, p. 11). This requires a 'strong contraction' of industries including fossil fuels, industrial livestock production and speculative finance, offset by long-term investment in renewable energy, public transport, commons-based circular manufacturing and retrofitted buildings. Raworth suggests we invest in sources of genuine wealth, 'natural, human, social, cultural and physical from which all value flows, whether it is monetised or not'. This will open opportunities for 'the roles of the market, the state and the commons as means for provisioning for our needs' (Raworth, 2018, p. 276). The roadmap to her vision of 'human prosperity in a flourishing web of life' includes becoming 'agnostic in the sense of designing an economy that promotes human prosperity whether GDP is going up, down or holding steady'.

Notions of a steady-state economy, one that does not exceed ecological limits, are 'difficult to envisage in an era where economic growth is seen as one of the most critical measures of political success' (Wright & Nyberg, 2015, p. 179). A selective degrowth calls for us to consider what production and consumption practices need to be stopped and which need to be encouraged (Kallis, 2011). This thinking sits uncomfortably with many of the 'promissory narratives' (Sexton, Garnett & Lorimer, 2019) spawned to generate investment capital and consumer interest in animal-free alternatives, including the plant-based proteins and cellular agriculture in the form of lab-cultured or 'clean' meats, milks and cheeses discussed in Chapter 3. Where do these fit in the ideal 'universal' diet that is heralded to bring about the large-scale transformation our food systems need?

THE DIET WARS

Today's responsible, ethical eater is bombarded with multiple framings of healthier bodies, food justice, animal welfare and climate-stable futures. Many of these focus on plant-based diets. Strong counter-narratives have emerged from the livestock sector across mainstream media, blogs, social media and public campaigns. These contestations came to a head in 2019, leading *The Observer* to declare diet the 'latest front in the culture wars' (Anthony, 2019). This statement was triggered by reactions to the publication of *Food in the Anthropocene: EAT-Lancet Commission on healthy and sustainable food systems* (Willet et al., 2019). It calls for a 'Great Food Transformation' focused primarily on the 'environmental sustainability of food production and health consequences of final consumption' (Willet et al., 2019, p. 448). Its 37 expert authors argue that eating more plant-based foods will lower rates of cancer and heart disease and result in more sustainable land use while reducing carbon emissions. Making recommendations at the global policy level, many of which are common sense, for example decreasing livestock production, improving governance of land and water and minimising food waste, it proposes a 'planetary health diet' high in plants and whole grains and low in animal and processed products. Unfortunately, even though it might be well-intentioned, the prescriptive nature of this diet, which effectively maps micro-nutrients across the food groups, fails to grapple with the barriers – cultural, socio-economic and otherwise – that might be faced by people implementing dietary changes. Essentially, it fails the relatability test in that it is not a diet that many people might realistically picture themselves embracing. It bears uncanny resemblance to the 'salmon and broccoli diet' that is now promoted in countries like Mexico to overcome the triple burden of over, under and malnutrition; a diet that has eclipsed the traditional 'holy trinity' of beans, maize and rice (Gálvez, 2018).

Critics have described the *EAT-Lancet* mission as technocratic, 'dangerously simplistic' and a campaign 'not so much to promote animal welfare as to open up for "Big Ag" lucrative new markets and feed the hunger of governments for new tax bases' (Leroy & Cohen, 2019). Most notable is the apparent indifference of the *EAT-Lancet* report to the political economy of our food systems and the ongoing production of inequality which has its roots, in many countries, in racialised land and labour relationships. Reckoning with these inequalities is avoided in a system where noncommunicable diseases (NCDs) are viewed as a result of poor decisions and food behaviours, participation in the alternative food system is equated with transformation of that system, and personal self-care is conflated with 'good citizenship'. Describing the report as 'a top-down attempt by a small, unrepresentative, dogmatic global elite to mould public agriculture policy' Joanna Blythman (2019b) minces (pardon the pun) no words:

> *We've lost the plot when arcane imports and genetically modified fake meat burgers dreamed up by venture capitalists in Silicon Valley are portrayed as more acceptable than a lamb chop from a British hillside. But as part of its 'Great Food Transformation',* Eat-Lancet, *and people like George Monbiot, actively want to stamp out the existing multiplicity of distinctive and diverse food*

cultures that are predicated on local history, seasons, traditions, cultivars, breeds, and artisan methods, and they want to replace them with a monocultural, globalised diet, one that's centered on factory and laboratory food. They would replace this culinary richness and natural biodiversity with a top-down, 'we know what's best for you' universal diet.

(Blythman, 2020b)

The *EAT-Lancet* report triggered a 'digital backlash' in the form of a countermovement. The *Lancet* published a social media analysis of the ensuing conversation 'to show how a rapidly changing media landscape and polarisation pose serious challenges to science communication on health and climate issues' (Garcia, Galaz, & Daume, 2019, p. 2154). Promoting the hashtag #yes2meat a few days before and then vigorously after the launch of the report, the movement gained sufficient strength to outnumber positive or neutral tweets under the official #EATLancet hashtag by 10 to one. The *Lancet* accuses this 'new sceptical online community' of 'intentional dissemination of misleading content', content 'pollution' and disinformation, and notes the need for pro-active avoidance of 'manipulation and misinformation about issues of fundamental importance for human health and the planet'.

The *Eat-Lancet* universal diet is but the latest of a series of high-minded, trenchant critiques of the industrial food system from a WEIRD, middle-class alternative food movement (AFM) that promotes 'voting with our forks'. A second wave of this movement, emerging since 2010 but with roots in civil rights struggles, is food justice,

…an approach to critiquing the food system that rested on expressly anti-racist and class-conscious

principles, and that foregrounded the leadership of
those most adversely affected by the industrial food
system.

(Povitz, 2019, p. 240)

Food justice presents 'productive openings…to explore
how power nuances the role of voice, storytelling and posi-
tionality in narrations of food-related problems and their
solutions' (Gordon & Hunt, 2018, p. 14). It critiques the
homogeneity and whiteness of AFMs (Alkon & Agyemanm,
2011; Guthman, 2008). Sustainable eating practices and lifestyle movements including veganism have been subject to the
same critique. Some vegans have been accused of being blind
to the fact 'that the very category of human is not universal,
but rather just as contentious as the category of animal'
(Polish, 2016, p. 377). Missteps include PETA's tone deaf 'Are
animals the new slaves?' exhibition; the twitter campaign
#AllLionsMatter, invoked after the much-publicised slaughter
of Cecil the lion; and the 2014 publication of *Thug Kitchen*, a
vegan recipe book written in 'verbal blackface' by white
authors. Jennifer Polish notes that, collectively, these examples
highlight

…the ways that white veganism often enacts upon
the bodies of people of colour the same thoughtless
devaluation that they accuse others of when they call
for people to recognise that animal lives do, in fact,
matter.

(Polish, 2016, p. 386)

But there is no universal 'veganism', any more than there is
one way of being omnivorous. Vegan protest can be 'anti-
speciesist, anti-racist, environmental or health-centric'; vegans
of colour may choose to use their veganism 'as a tool to
decolonise the body from a colonial diet that is killing the

black community' (Greenebaum, 2018, p. 682). As a '*counter*cuisine', veganism embraces a wide variety of meanings and interpretations, and like all social movements, it experiences 'regular negotiation of conflicts internal to the movement as well as those associated with outside entities in the wider social movement environment' (Wrenn, 2019, p. 190; see also Tsing, 2005).

Our own sensitivities to the stereotyping of our food practices, and our food politics, give us fair warning that other people's diets are none of our business. How can 'we' determine a suitable, let alone universal, diet for humans from diverse cultural and ethnic groups with foodways ranging from the carnivorous diets of those who live in the coniferous forests and tundra of the Arctic North to the herbivorous diets of many tropical and desert peoples, the basic, unprocessed diets of traditional societies and the elaborate cuisines in between these? Dietary habits transfer between generations; they are deeply embedded in culture. Flexibility is the key to thriving in a changing climate where our foodscapes are going to change profoundly.

The FAO's 2020 State of Food Security and Nutrition (SOFI) report notes that meeting the SDG 2 target of Zero Hunger by 2030 is only possible if people have enough food to eat and if what they are eating is 'nutritious and affordable' (FAO, 2020b). It also wades into the diet wars, analysing the health and climate costs of current consumption patterns. Yet, according to right to food watchdog FIAN International, it neglects to address 'the social costs of the dominant industrial food system including land grabbing, labour exploitation and gender inequality'. Thus, it sidesteps

> *...the structural causes of the unaffordability of healthy diets, such as the colonial heritage of the*

> *global division in food production or policies that*
> *promote export-oriented staple cash crops at the*
> *expense of healthy, diverse and traditional food for*
> *domestic consumption.*
>
> (FIAN International, 2020b)

Changing our diets might enable us to reconcile with our own consciences, but this is not enough to reckon with the fraught politics of a food system committed to over-production, the maximisation of profits and the concealment of injustices, against our own bodies, others and the planet.

(RE)FRAMING THE PROBLEM

The critical flaws in the idea that we can 'change the world one meal at a time' become clear when we apply a political ecology lens to the contexts in which our decisions about food are made. Julie Guthman stresses that to make political choices that matter in our foodways 'requires much more attention to the broader injustices that the cheap food dilemma rests on and perhaps less attention to what's on the menu' (Guthman, 2011, p. 194). She provides a compelling argument that corporate profiteering and state regulation (or lack thereof) that increases our exposure to toxins and chemicals plays a significant role in decisions about diets and health ways. She condemns the 'neoliberal ideologies' of individual responsibility and personal choice that let capitalism, in its failure to address harmful externalities of production, off the hook.

These ideologies pervade the Age of Nutritionism (Scrinis, 2013) which has proved a boon to the food industry in terms of new markets for faddish food products and lifestyle behaviours. The alternative food movement originated in opposition to the ills of the conventional food system yet has contributed to the construction of health as a discursive marker. Deviating from a set of food provisioning activities that model social forms of organisation and governance such as cooperatives and Community Supported Agriculture (CSA), today's alternative food movement has become, according to Guthman, 'less concerned with using food practices in the service of social change than with changing the food itself' (2011, p. 142). To eat at the farmers' market (local, seasonal, organic) and engage in conviviality (which literally means 'living together', a theme of the Slow Food movement) is to engage in a form of 'lifestylism', where 'disease prevention morphs into moral prescriptions as to how one should live' (p. 147).

> *The idea that a food system can be transformed by selling and buying good food (through informed choice) is a huge concession to the neoliberal idolatry of the market.*
>
> (Guthman, 2011, p. 148)

Illusions of purity in our foodways are confounded by our entanglement with other actors in our food practices. These are embedded within wider food and energy systems in which we are all differently situated. Take examples of free-range eggs, wild-caught fish and grass-fed meat which are more expensive than factory-farmed animal-based food. If we 'hold ethics to the level of the individual, we restrict ethical choice to those who are most privileged by and within the system', those who can access and afford these products (Shotwell, 2016, p. 125).

The same tension applies to those who grow our food. Expensive and prescriptive organic certification programs are a kind of incentive-based regulation based on the idea that those who want food free of pesticides should pay for it. They favour well-capitalised producers and elevate the prices of certified produce beyond the reach of everyday eaters, creating 'ethical enclaves' (Toit, 2001). But they offer well-heeled consumers reassurance against food scares and provide an easy way of achieving this. As such the alternative food movement's focus on local food and organics, for example, 'abandons' regulatory reform in conventional agriculture. Jill Lindsey Harrison, in researching cases of pesticide drift in California, observes that pesticide politics have shifted since the 1960s 'from collective action that crossed class divides and interrogated injustices at the site of production, to individualised consumer politics whose environmental benefits are tied to shoppers' purchases' (Harrison, 2011, p. 158). Regulation through market choice rather than regulation neglects the ills of a broader supply chain in which farmers, agricultural labourers and food service workers cannot afford the the very products they provide, and is therefore a failure as a theory of change. Food chain workers are constrained in improving the quality of their diets through lack of time, money and access (McMillan, 2012). For these essential workers, 'bringing good food to others isn't changing the conditions of exploitation and oppression or addressing the privilege that also results from pervasive inequality' (Guthman, 2011, p. 161). Therefore, though it is a tall order,

> *...to achieve effective, just solutions to today's environmental problems, we must account for social injustices – inequalities, oppression, a lack of participatory parity and inadequate basic capabilities – and actively work to rectify them.*
>
> (Harrison, 2011, p. 204)

So-called 'common sense' explanations of global warming and obesity are 'apolitical' in that they fail to acknowledge or account for the role of power in either 'producing environmental changes or defining them as problems'. It is time we accepted that 'social, cultural and political-economic relations profoundly affect both the materiality of the biophysical world and our understanding of it' (Guthman, 2011, p. 9).

CONFRONTING FOOD APARTHEID

Framing both how we eat and how our activities contribute to global warming as *problems of human behaviour* fails to recognise that we eat, shop and live in a food system. It relies on a model of food consumption as a 'predominantly cerebral and rational activity, in which consumers exercise their choices by carefully evaluating the different options available to them in the marketplace' (Evans & Miele, 2017, p. 236). Anyone who has ever visited the supermarket knows this is not actually how we operate. In reality, an imposing barrier to changing food practices is the fact that they are habitual and social, rather than rational and cognitive. The 'attitudes, behaviour, choice', or ABC model, fails to address the ways in which we live, or think we need to live. Framing citizens as consumers and placing responsibility on individuals 'obscures the extent to which governments sustain unsustainable economic institutions and ways of life, and the extent to which they have a hand in structuring options and possibilities' (Shove, 2010, p. 1724). The many half-conscious, routinised actions we make in everyday life are shaped by socio-technical structures. Viewed from this perspective, consumer practices are resistant to change because they are interwoven into other practices – like visiting the supermarket between work and school. Most importantly, we need to recognise that they are restricted by discriminating policies that create what Devita Davison, Executive Director of FoodLab Detroit, describes as 'food apartheid':

We don't use the words 'food desert'. What we use is a more appropriate term, 'food apartheid', meaning our neighbourhoods and communities in the city of Detroit and communities that have been occupied with Black and brown bodies all over the country, whether it's Detroit, Harlem, the Bronx, Oakland, parts of Baltimore, DC – we live under food apartheid. A desert is a natural phenomenon, but having lack of access to fresh, healthy, affordable food is not natural, nor is it accidental.

(Davison & Kinsman, 2020)

We seldom make decisions on the basis of genuine free will; rather we are shackled by structural conditions including the organisation of society in time and space. Given that long and complex food supply chains distance us from the origins of our food, we need a high level of food and health literacy to make educated choices where we weigh short-term benefits or satisfaction against long-term health consequences. It also assumes transparency in regard to how our food is produced.

Getting accurate and unbiased information about industrial practices is challenging in modern society where 'distance and concealment operate as mechanisms of power' (Pachirat, 2013, p. 3). Practices of industrial meat production, as discussed, are 'hidden in plain sight' because the realities of mass killing 'antagonise "civilised" sensibilities'; the fact is we can't handle the truth (p. 251). Workers, like the animals they handle, are subordinated by their role as disposable commodities in meat processing and packaging plants. In the US state of Georgia, the self-declared 'poultry capital of the world', competitive advantage is achieved through low unionisation rates and the lowest average of all hourly poultry processing wages at US$9.45/hour (Freshour, 2019, p. 123).

Worker organising, which dates back to 1979 in official poultry industry records, was initially led by 'mostly Black women', who represented 'those getting banged around the most' (p. 127). In response to strikes, management hired undocumented Hispanic workers in a deliberate act of labour displacement. When migrant and Black workers created coalitions and mass mobilisations triggered the closure of 22 plants owned by Tyson, Perdue and Gold Kist in 2006, the industry activated a 'gendered racial removal program' to divide and conquer opposition. Undocumented workers were fired, and the workforce is again primarily female and Black and engaged in political struggle against plans such as those to deregulate line speeds to meet Chinese processing rates of 225 birds per minute, up from 140 (Freshour, 2019, p. 123). Worker safety, food safety, animal welfare and inspection processes are now in the spotlight with COVID-19 crippling not just poultry but beef and pork plants nationally as the meat industry bleats that it is vital to the nation's food security.

Unfortunately, those of us engaging in 'small-p politics' by voting with our wallets and our forks to whittle away at the profits of Big Food are not going to stop these violations. Individual action will never be enough to drive the transformative change needed to respond to the challenges of our food system if we fail to immediately contest the inherent structural violence in our foodways, the 'social arrangements that put individuals and populations in harm's way' (Farmer, Nizeye, Stulac, & Keshavjee, 2006, p. 1686). Unevenly distributed across populations in accordance with vulnerabilities, structural violence reveals 'symptoms of deeper pathologies of power' directly related to social conditions that put them at risk of harm (Farmer, 2004). As long as we subscribe to the ABC model where we explain chronic, diet-related diseases to be rooted in individual knowledge and behaviour, we will continue to think they fall beyond the orbit of politics and economic development (Gálvez, 2018, p. 6). Therefore, the state of our health owes considerably more to a failure of political will than our lack of personal will power.

Further, we are running out of time. Climate change is going to impact our diets before we make a dent in global emissions through our purchasing power. Here, as political consumers, we are indulging in time discounting, that tendency to 'value a reward to a lesser degree the farther in the future it is received' (Barlow, Reeves, McKee, Galea, & Stuckler, 2016, p. 810).

This short-term thinking supports a Global Syndemic of obesity, undernutrition and climate change driven by the food and agriculture system, transportation, urban design and land use (Swinburn et al., 2019). There are clear actions that would work to mitigate the impacts of two or three of these factors simultaneously yet in many countries efforts to take measures – such as integrating genuine sustainability principles within their dietary guidelines – are stymied by 'strong food industry lobbies, especially the beef, dairy, sugar, and ultra-processed food and beverage industry sectors' (Swinburn et al., 2019, p. 792). Partnerships between Big Food and governments are a serious risk to public health as in the case of the sugary drinks sector, which spent close to US$50 million in 2016–2017 to lobby against government-led initiatives to reduce soda consumption. Expansion of markets into the Global South is largely unregulated in and around schools in emerging markets like Nepal, Ghana, South Africa and Mongolia. Like Big Polluters, multinational agribusiness exercises immense obstructive power in governance arrangements that legitimise industry engagement in public policy; in effect, loose regulation is a 'huge subsidy to the food industry' (Guthman, 2011, p. 129). The 'revolving door' between corporate and government agencies protects private interests at the expense of human and environmental health (Nestle, 2007). The former UN Special Rapporteur for the Right to Food, Hilal Elver, reports global corporations manufacturing pesticides are guilty of 'systematic denial of harms' and 'aggressive, unethical marketing tactics'. She condemns lobbying practices that have 'obstructed reforms and paralysed pesticide restrictions'. Corporate elites infiltrate federal regulatory agencies and 'cultivate strategic public-private partnerships that call into question their culpability or help bolster the companies' credibility' (Elver, 2017, p. 19). This credibility is propped up by networks of academics and regulators recruited as consultants.

RESISTING THE FORCES OF INACTION

On the one hand, COVID-19 is a critical focusing event, and an opportunity for a re-visioning of our food systems, but it is also serving as a distraction from the 'other emergencies that are destroying our foundations' (Shearman, 2020). Hunger will be more deadly than COVID-19 to the poor.

The World Food Programme (WFP) already provides food to millions of people living in endemic hunger and poverty. The economic consequences of the pandemic will be more devastating than the actual virus, particularly in Africa's highly export-dependent Angola, Nigeria and Chad and highly import-dependent Somalia and South Sudan. The Middle East, already crippled by conflict, sees Yemen, Iran, Iraq, Lebanon and Syria all facing severe economic problems (WFP, 2020). In WEIRD societies, the food bank nations (Riches, 2018), the pandemic highlights the unsustainability and injustice of community reliance on a charity or food bank social service model within which 'service provision maintains rather than challenges unequal power relations' (Povitz, 2019, p. 13). In many ways, the anti-hunger movement has sought to feed the poor, rather than give them power, while Big Food benefits from a corporate charity model that reduces dumping costs and provides tax write-offs,profit protection and great public relations. Walmart, well-known for its below-market wages, has donated over US$2 billion in grants and food to charities since 2010. Feeding America, the national representative of food banks in the US, has partnered with Monsanto in its 'Invest an Acre' program which invites farmers to donate proceeds from an acre of production to their local food bank while Archer Daniels Midland and Monsanto match the funds (Fisher, 2020). De-politicising hunger in this way reduces the collective power of labour, anti-hunger and alternative food movements. In approaches which emphasise social leadership and solidarity relationships,

> ...*people work for change in a way that recognises their interdependence: it emphasises the responsibilities, interests or identity aspects that are shared, so as to erase or lighten differences. Charity, by contrast, emphasises and preserves the differences between givers and receivers.*

(Povitz, 2019, p. 9)

Actively dismantling racism in our societies means addressing disparities across the entire food supply chain. In the US food service, workers of colour are concentrated in the lowest paid jobs where they are subjected to wage theft, dangerous working conditions, discrimination and harass-ment. Women, undocumented workers and those exploited through modern day slavery are among the most vulnerable (Nolan & Boersma, 2019). Yet many of these workers are 'taking on Goliath' in unions and organised actions that are successfully challenging corporate power through protest, litigation, coalition building and public education (Araby, 2020; Jayaraman, 2020; Jayaraman & De Master, 2020; Walter, 2012). They demonstrate 'we can create our own future, even under economic and political conditions not of our choosing' (Araby, 2020, p. 171). Leading the charge is the Coalition of Immokalee Workers (CIW), based in Florida's southern food basket. This movement of thousands of Latinx, Mayan Indian and Haitian immigrant fruit pickers is confronting decades of poverty and abuse in the tomato industry through effective nationwide activism targeting fast-food restaurants. Since its formation in 1993, the CIW has triggered nine federal prosecutions freeing over 1,200 forced labourers, but its boldest turn came in 2001 with a nation-wide farmworker boycott calling Taco Bell to account for human rights abuses in the fields where its produce is grown and picked. Four years of relentless campaigning cultivated enough community support to force the fast-food giant to agree to improve the wages and conditions of workers along the supply chain. Capitalising on this landmark success, the movement consolidated the Alliance for Fair Food (AFF), creating and expanding the 'Fair Food Principles' set in the Taco Bell agreement with McDonald's in 2007. Soon after, Burger King, Subway and Whole Foods signed the AFF agreement. Food service providers including Sodexo and Bon Appètit followed. In 2010, in what was described by the *New York Times* as 'possibly the most successful labour action in the US in 20 years', the Florida Tomato Growers Exchange extended the AFF's Fair Food Principles – codes of conduct, health and safety initiatives, fair complaint resolution systems and education programs – to 90% of the industry. In 2017, *Harvard Business Review* declared CIW's achievements 'a David-versus-Goliath triumph...a scalable model and leverage point to gain humane working conditions and a 70% increase in wages' (Wolf Ditkoff & Grindle, 2017). These struggles, over years and at great expense, are led by those who, with the health workers around the world, have carried us through the COVID-19 crisis. They reckon with the hard truth that we cannot change the food system without

> *...actively battling corporate agriculture and without ensuring the social dimensions of alternative agriculture – small-scale production, family or communal ownership, community solidarity between farmers and consumers, short distance from farm to table.*

> (Bello, 2009, p. 144)

Finally and vitally is the urgent need to decarbonise our food systems and our wider economy. When it comes to climate change 'the forces of inaction have deep pockets' (Nixon, 2013, p. 39). In no other industry is more capital invested than oil and gas – the total market capitalisation grew 186% to US\$3,153 trillion between 2001 and 2010. Add in state-owned and the total reaches US\$6,729 trillion, more than the world's banks combined (Angus, 2016, p. 170). Capitalism and fossil fuels are so tight that is hard to imagine one with the other. The possibility of a full transition to renewables has been compared to the Apollo moon project and the US Interstate Highway System by Stanford and University of California-Davis researchers, who proclaim

> *...we really need to just decide collectively that this is the direction we want to head as a society...the biggest obstacles are social and political – what you need is the will to do it.*

> (Bergeron, 2011; Delucchi & Jacobson, 2011)

The European Union is taking steps with a circular bio-economy approach including soil regeneration and agro-ecological approaches and transitions in waste management such as the Italian Composting and Biogas composting plants, *Consorzio Italiano Compostatori* (CIC). Encouraging huge investments in new technologies but also social innovation by active citizens, the EU is recognising climate change is the 'biggest market failure of our times' that can 'only be corrected through the intervention of governments and by steering zero-carbon innovation'. This 'challenging transition', it adds, must be led by 'a new generation of brave and visionary actors' (European Commission, 2018, p. 167).

We can be assured that the fossil fuel and food lobbies will hope we are distracted from the hyperobject by the double blow of pandemic and global recession. They will be relying on that familiar 'quirk of human psychology: that we react more strongly to an imminent threat than to a distant one' (Seccombe, 2020, p. 8). This thinking aligns neatly with conventional business models, 'beholden to the bottom line and short-term thinking', on a time scale that 'privileges the present, which is profoundly out of sync with environmental realities and democracy itself' (Taylor, 2019, p. 295). A case in point: as we endure the second wave of the pandemic, parliament is suspended, and Australian state governments have 'slipped thorough' several decisions to expand fossil fuel activities, for example, coal mining under the Woronora reservoir, one of Sydney's water sources, and the Shell-operated joint venture Surat gas project, located in one of Queensland's fertile foodbowls (Brett, 2020). Excavation work has commenced on the Adani Carmichael coal mine, labelled 'the world's most insane energy project' by *Rolling Stone* (Goodell, 2019). When Greenpeace revealed the scale of the Adani project in 2012, nine mines set to produce up to 330 million tonnes of coal a year, twice annual exports at the time, Trade Minister Craig

Emerson accused supporters of the switch from coal to renewable energy of delusional thinking which would lead to 'a global depression' and 'mass starvation'. Prime Minister of the moment, Julia Gillard, reassured the industry that coal had a 'great future' in Australia in alignment with her rival Tony Abbott's infamous declaration that coal is 'good for humanity' (Bradley, 2019). Adani has an appalling environmental record in India but was deemed a 'suitable operator' under Queensland's Environmental Protection Act. Concerns about impacts on the Carmichael River and groundwater, raised by the Commonwealth's Independent Expert Scientific Committee on Coal Seam Gas and Large Coal Mining Development, were ignored by the state government. Unsurprisingly, considering that in 2010, under Labour Premier Anna Bligh, it launched *CoalPlan 2030*, a proposal to double the state's coal production to 340 million tonnes a year over the next two decades (Queensland Government, 2010). Claims of economic benefits, including 10,000 jobs – in an area where unemployment is around 9% – and AU$22 billion in tax revenue, are highly contested. These promises are part of the enduring myth of the coal industry and its contribution to the lives of ordinary Australians – '#coalisamazing!' spruiked a 2015 industry campaign. Corporate donations, lobbying and political favours are endemic in the 'game of mates', better understood as a form of regulatory capture or 'government failure in which policy decisions are shaped not by the public interest but by interest groups or industries' (Bradley, 2019). Meanwhile, we are crippled by bipartisan inaction on climate change; there are no short-term targets to reach net zero emissions by 2050, and both sides of politics support carbon capture and storage technology 'as a sop to fossil-fuel interests' (O'Malley & Foley, 2020).

Our trade in unprocessed commodities has made us a rich country. Yet like other 'resource cursed' (Brett, 2020) nations such as Nigeria, Angola and Venezuela, we have become dependent on export income, complacent about investment in domestic manufacturing, and vulnerable to exploitation by political elites. As the world's second largest exporter of coal and the largest exporter of liquefied natural gas (LNG), we are now the world's largest exporter of fossil fuels behind Saudi Arabia and Russia. But capital is deserting fossil fuels – at time of writing Adani has been unable to find a single bank willing to fund the Carmichael mine (Brett, 2020). Risk is driving fund managers and nervous insurers away from coal projects. Public sentiment has been turned by mobilisation of groups including Greenpeace, the Australian Youth Climate Coalition and the #StopAdani movement, groups represented, variously, as 'economic saboteurs', vigilantes and terrorists in a pervasive political discourse where 'clean coal' is obtainable through 'sustainable mining' and 'new megamines are desirable, inevitable and essential to maintain Australia's national destiny' (Brevini & Woronov, 2017). This rhetoric replaces meaningful political engagement and exposes hollow diplomacy.

To escape this lock-in, we need to invest in renewable energy, the potential of which is clear. Australia's climate is naturally advantageous to the production of solar energy, for example, which could re-boot manufacturing and create jobs in regional areas (Garnaut, 2019). South Australia is already leading the way in using solar power and desalinated seawater in agriculture. Port Augusta's Sundrop Farms grows tomatoes across 49 acres while producing the equivalent of 180 Olympic swimming pools of water and displacing more than 2 million litres of diesel each year (Falzon, 2016). This is a smart alternative to the extraction of transition fuels, which serve as 'a layer of icing on top of a bigger and bigger piece of energy cake that we eat each day' (Jahren, 2020, p. 114). The Lock the Gate Alliance is leading local community opposition to coal seam gas fracking. It calls on landowners to shut their gates to gas companies, declaring its mission 'to protect Australia's natural, cultural and agricultural resources from inappropriate mining and to educate and empower all Australians to demand sustainable solutions to food and energy production' (Lock the Gate, n.d). These types of grassroots, direct actions against fossil fuel projects, collectively described as Blockadia, now represent a 'roving transnational conflict zone' (Hawken, 2007; Klein, 2014) driven by ordinary people frustrated with the lack of leadership on climate change. Many of these movements, pitted against developments such as the Dakota Access Pipeline, Kinder Morgan, Keystone XL, Enbridge Line 9, Bayou Bridge and TransCanada Energy East in North America, connect First Nations peoples and frontline communities in powerful coalitions (Estes, 2019). They are complemented by divestment campaigns which are expanding across university sectors to pension funds, municipal governments and philanthropic organisations. The logic of divestment 'couldn't be simpler; if it's wrong to wreck the climate then it is wrong to profit from that wreckage' (McKibben cited in Wright & Nyberg, 2015, p. 182). Divestment might not bankrupt fossil fuel companies but it will make their social licence to operate less secure. The fastest path to decarbonisation is making it clear that climate change is an economic risk, and that mitigating it is essential to our survival as a species.

Faced with this ultimatum, we need to be doubly alert to the clarion call of the 'Good Anthropocene'. This narrative seduces with the promise of a 'corporate environmentalism' that presents profit-hungry TNCs as partners in the mitigation of global warming. Further, it promotes their power to solve our warming woes with technological innovation, including a lavish suite of novel foods and climate-smart agricultural products that open new profit channels for Big Food and agro-industrial seed conglomerates (McMichael, 2009). For those of us with privilege and a full pantry, resistance means 'refusing the smugness of "green" consumption' (Wright & Nyberg, 2015, p. 194) and other dimensions of capitalism's 'green bubble' (Morton, 2010, p. 100). It also means refusing to 'recalibrate our expectations' so that our warming condition, with all its discomfits and injustices, becomes acceptable (Wallace-Wells, 2019, p. 236). Let the 'anticipatory mourning' (DeLoughrey, 2019, p. 4) begin but not render us immobile.

For a democratic food system, like a democratic world, 'is not a trickle-down proposition. Rather it foments up. Impolitely. It is nothing short of a rebellion' (Jayaraman & De Master, 2020, p. 4). We should be 'enraged' against the violence of discriminatory policies that constrain women's autonomy in agriculture and life; against the hegemonic capitalist paradigm that exploits lives and nature; and against the invisibility of those shaped 'not *just* by their gender but also by their race, ethnicity, caste, class, sexual orientation or identity, global location ...and (dis) ability' (Andrews, Smith, & Morena, 2019, p. 8). Food, in all its materiality, embodiment and embeddedness demands our attentions; it 'encourages the "right to the visceral, spiritual and sensory freedoms" as well as the right to outrage, revolt and anger' (cited in Andrews et al., 2019, p. 9).

A DISH WITH ONE SPOON

A cultural revolution or *pachakuti* – profound overhaul of the social order – is necessary to dismantle our irrational food system. This is more than a socio-technical transition across all dimensions of life – the technological, material, organisational, institutional, political, economic and socio-cultural

(Markard, Raven, & Truffer, 2012). Through a socio-technical lens, a food system is made of people, inter-connected actors and elements, institutions and knowledges. But perhaps it is better understood from an Indigenous point of view as a 'more than-human-society' (Watts, 2013) where ethical frameworks and inter-species treaties and agreements further our ability to respect those with whom we share ter-ritories and owe responsibilities.

The Indigenous peoples of the Great Lakes region in North America speak of 'a dish with one spoon' or *Gdoo-naaganinaa* to describe how the land can be shared for the mutual benefit of all its inhabitants. This concept united the mighty Iroquois Confederacy or "Great League of Peace" made up of Seneca, Cayuga, Oneida, Onondaga and Mohawk Nations and was the basis of a treaty between the Anishinaabe and Haudeno-saunee people. *Gdoo-naaganinaa* places on all parties of an agreement 'responsibility to ensure that the dish would never be empty by taking care of the land and all of the living beings on it' (Duhamel, 2018). The absence of a knife in this table setting reflects to the need to maintain peace for the benefit of all, united in a kinship that extends beyond blood relations to 'our cousins' or *kiciwamanawak,* a term Cree Elders use to explain how treaties are 'adoptions of one nation by another' (Johnson, 2007, p. 13). In this relationship Cree and settlers are equal, 'two families living in the same territory' (Johnson, 2007, p. 20). Montreal Lake Cree Nation member Harold Johnson says 'if we return to the original intention of treaty and recognise that we are relatives, *kiciwamanawak*, we should be able to walk into the future in a good way' (John-son, 2007, p. 121). 'Making kin is to make people into familiars in order to relate', writes Dakota scholar Kim Tall-Bear (cited in Estes, 2019, p. 256). This ability to relate is essential to the processes of decommodification, co-operation and social production that open up possibilities for an economic agenda that 'respects the limits of local and global ecosystems, including the realities of climate change' (Hilary, 2013, p. 159).

However you imagine the future of our foodways, we need political and aesthetic strategies to re-vision a new food system that 'seeks to resurrect and then sustain what we have seemingly lost' in a radical plan – radical in that it presents 'a vision of the future that is achievable while being very much at odds with the way the world currently works' (Akram-Lodhi, 2013, p. 169). We need to be transition activists, promoting values including militancy, humility, independence, defiance and creativity, presenting pragmatic solutions and pressuring states to work with broad coalitions of food citizens. We need critical, provocative and experimental approaches that break out of the Fordist thinking that dominates industrial food with its processes of replication, repetition and commodification that make food an object when it should be a relationship, the most intimate one we will ever share.

REFERENCES

ABARES. (2019). Snapshot of Australian water markets. Retrieved from https://www.agriculture.gov.au/abares/publications/insights/snapshot-of-australian-water-markets#australian–water-markets-why-where-who-and-how

AbCF. (2020). Carbon farming: Core benefits. *Aboriginal Carbon Foundation*. Retrieved from https://www.abcfoundation.org.au/

Abrol, D. (2019). Concentration in global seed and agro-chemical industry: Implications for Indian agriculture. In L. Premkumar (Ed.), *Corporate concentration in agriculture and food*. Bangalore: Focus on the Global South.

Ackerman-Leist, P. (2013). *Rebuilding the foodshed: How to create local, sustainable and secure food systems*. White River Junction, VT: Chelsea Green Publishing.

Aither. (2019). *Water supply and demand in the southern Murray-Darling Basin*. Retrieved from https://waterregister.vic.gov.au/images/documents/Water-Supply-and-Demand-Report_Aither_FINAL.pdf

Akram-Lodhi, A. H. (2013). *Hungry for change: Farmers, food justice, and the agrarian question*. Black Point, NS: Fernwood Publishing.

Alaimo, S. (2017). Your shell on acid: Material immersion, anthropocene dissolves. In R. Grusin (Ed.), *Anthropocene feminism*. Minneapolis, MN: University of Minnesota Press.

Alexander, M. (2012). *The new Jim Crow: Mass incarceration in the age of color blindness*. New York, NY: The New Press.

Alkon, A. H., & Agyeman, J. (2011). *Cultivating food justice: Race, class, and sustainability*. Cambridge, MA: MIT Press.

Alkon, A. H., Bowen, S., Kato, Y., & Young, K. A. (2020). Unequally vulnerable: A food justice approach to racial disparities in COVID-19 cases. *Agriculture and Human Values*, 1–2.

Anastario, M. (2019). *Parcels: Memories of Salvadoran migration*. New Brunswick, NJ: Rutgers University Press.

Andrews, D., Smith, K., & Morena, M. A. (2019). Enraged: Women and nature. In *Right to food and nutrition watch: Women's power in food struggles*. 11, Heidelberg: FIAN International & Brot für die Welt.

Angus, I. (2016). *Facing the Anthropocene: Fossil capitalism and the crisis of the Earth system*. New York, NY: Monthly Review Press.

Anthony, A. (2019). How diet became the latest front in the culture wars. *The Observer*. Retrieved from https://www.theguardian.com/environment/2019/mar/17/how-diet-latest-front-culture-wars-eat-less-meat-lancet

Araby, J. (2020). Progress over poverty through political power. In S. Jayaraman & K. De Master (Eds.), *Bite back: People taking on corporate food and winning* (pp. 159–174). Oakland, CA: University of California Press.

Arantini, L. (2020). South Dakota pork plant closes after over 200 workers contract COVID-19. *The Guardian*. Retrieved from https://www.theguardian.com/us-news/2020/apr/13/ south-dakota-pork-plant-closes-after-200-workers-contract-covid-19

Arendt, H. (1958). *The human condition*. Chicago, IL: Chicago University Press.

Askew, K. (2019). Introducing the NaSu burger: The basic idea is totally different to Impossible or Beyond. *FoodNavigator-USA*. Retrieved from https://www.foodnavigator.com/Article/2019/ 10/30/Vegan-clean-label-and-sustainable-Introducing-the-NaSu-Burger?utm_source=newsletter_daily&utm_medium=email& utm_campaign=04-Nov-2019&c=16HhEPChu5DjA%2FMqV ed360jR6JSAYtW2&p2=

Bailey, C., & Tran, N. (2019). Aquatic CAFOs: Aquaculture and the future of seafood production. In B. Winders & E. Ransom (Eds.), *Global meat* (pp. 55–74). Cambridge, MA: Massachusetts Institute of Technology.

Banwell, C., Broom, D., Davies, A., & Dixon, J. (2012). Restoring coherence to a stressed social system. In C. Banwell, D. Broom, A. Davies, & J. Dixon (Eds.), *Weight of modernity: An intergenerational study of the rise of obesity* (pp. 173–189). New York: Springer.

Barber, D. (2014). *The third plate: Field notes on the future of food*. New York: Penguin Press.

Barlow, P., Reeves, A., McKee, M., Galea, G., & Stuckler, D. (2016). Unhealthy diets, obesity and time discounting: A systematic literature review and network analysis. *Obesity Reviews, 17*, 810–819.

Barona, E., Ramankutty, N., Hyman, G., & Coomes, O. T. (2010). The role of pasture and soybean in deforestation of the

Brazilian Amazon. *Environmental Research Letters*, 5(2), 024002.

Baumann, S., & Johnston, J. (2010). *Foodies: Democracy and distinction in the gourmet foodscape*. Abingdon: Routledge.

BBC. (2013). World's first lab-grown burger is eaten in London. Retrieved from https://www.bbc.com/news/science-environment-23576143

Beaumont, P. (2020). 'Millions hang by a thread': Extreme global hunger compounded by Covid-19. *The Guardian*. Retrieved from https://www.theguardian.com/global-development/2020/apr/21/millions-hang-by-a-thread-extreme-global-hunger-compounded-by-covid-19-coronavirus

Behrendt, L. (2016). *Finding Eliza: Power and colonial storytelling*. Brisbane, QLD: University of Queensland Press.

Bell, J. D. & Ganachaud, A. (2013). Mixed responses of tropical Pacific fisheries and aquaculture to climate change. *Nature Climate Change*, 3, 591–599.

Bello, W., & Baviera, M. (2010). Food wars. In F. Magdoff & B. Tokar (Eds.), *Agriculture and food in crisis: Conflict, resistance and renewal* (pp. 33–50). New York, NY: Monthly Review Press.

Bello, W. (2009). *The food wars*. London: Verso.

Bendell, J. (2018). Deep adaptation: A map for navigating climate tragedy. Occasional Paper. Institute for Leadership and Sustainability (IFLAS). Retrieved from http://insight.cumbria.ac.uk/id/eprint/4166/1/Bendell_DeepAdaptation.pdf

Benjamin, A., & McCallum, B. (2009). *A world without bees*. New York, NY: Pegasus Books.

Berger, J. (2007). *Hold everything dear: Dispatches on survival and resistance*. New York, NY: Pantheon.

Bergeron, L. (2011). The world can be powered by alternative energy, using today's technology, in 20-40 years, says Stanford research Mark Z. Jacobson. *Stanford News Service*. Retrieved from https://news.stanford.edu/pr/2011/pr-jacobson-world-energy-012611.html

Berry, W. (2017). *The world-ending fire*. London: Penguin.

Bladow, K. (2015). Milking it: The pastoral imaginary of California's (non)dairy farming. *Gatronomica, 15*(3), 9–17.

Bleakly, S., & Hayes, M. (2017). Algal proteins: Extraction, application, and challenges concerning production. *Foods, 6,* 33.

Bloomberg News. (2019). Impossible foods is creating plant-based pork as it eyes China. *Bloomberg*. Retrieved from https://www.bloomberg.com/news/articles/2019-11-06/impossible-foods-is-creating-plant-based-pork-as-it-eyes-china

Blythman, J. (2015). *Swallow this: Serving up the food industry's darkest secrets*. London: Fourth Estate.

Blythman, J. (2019a). How vegan evangelists are propping up the ultra-procesessed food industry. *Mouthy Money*. Retrieved from https://www.mouthymoney.co.uk/how-vegan-evangelists-are-propping-up-the-ultra-processed-food-industry/

Blythman, J. (2019b). Why we should resist the vegan putsch. *The Grocer*. Retrieved from https://www.thegrocer.co.uk/consumer-trends/why-we-should-resist-the-vegan-putsch/575625.article

Blythman, J. (2020a). Veganuary is huge. But is it really as simple as animal foods bad, plant foods good?. *The Guardian*. Retrieved from https://www.theguardian.com/commentisfree/2020/jan/22/veganuary-animal-foods-plant-vegan

Blythman, J. (2020b). Why plant-based diets aren't credible. Retrieved from https://sustainablefoodtrust.org/articles/why-plant-based-diets-arent-credible-joanna-blythman-at-the-oxford-real-farming-conference/

Bonhommeau, S., Dubroca, L., LePape, O., Barde, J., Kaplan, D. M., Chassot, E., & Nieblas, A. E. (2013). Eating up the world's food web and the human trophic level. *Proceedings of the National Academy of Sciences*, *110*(51), 20617–20620.

Bonneuil, C., & Fressoz, J. (2017). *The shock of the Anthropocene*. London: Verso.

Bookchin, M. (1990). *Remaking society*. Montreal, QC: Black Rose Books.

Bové, J., & Dufour, F. O. (2001). *The world is not for sale: Farmers against junk food*. London: Verso.

Boychuk, R. (1992). The blue revolution. *New Internationalist*. Retrieved from https://newint.org/features/1992/08/05/blue

Bradley, J. (2019). How Australia's coal madness led to Adani. *The Monthly*, April. Retrieved from https://www.themonthly.com.au/issue/2019/april/1554037200/james-bradley/how-australia-s-coal-madness-led-adani#mtr

Brander, K. (2007). Global fish production and climate change. *Proceedings of the National Academy of Sciences Dec 2007*, *104*(50), 19709–19714, doi:10.1073/pnas.0702059104

Brett, J. (2020). *Quarterly Essay 78: The coal curse: Resources, climate and Australia's future*. Carlton, VIC: Black Inc.

Brevini, B., & Woronov, T. (2017). Nothing but truthiness: Adani and Co's post-truth push for the Carmichael mine. *The Conversation*. Retrieved from https://theconversation.com/nothing-but-truthiness-adani-and-cos-post-truth-push-for-the-carmichael-mine-85671

Brillat-Savarin, J. A. (1949). *The physiology of taste: Or mediations on transcendental gastronomy*. New York, NY: Vintage.

Broad, W. J. (2009). From deep Pacific, ugly and tasty, with a catch. *The New York Times.* Retrieved from https://archive. nytimes.com/www.nytimes.com/2009/09/10/science/10fish.html

Broad, G. M. (2016). *More than just food: Food justice and community change.* Oakland, CA: University of California Press.

Brown, L. (2011). *World on the edge: How to prevent environmental and economic collapse.* London: Earthscan.

Butler, J. (2005). *Giving an account of oneself.* New York, NY: Fordham University Press.

Cagle, S. (2020). 'A disastrous situation': Mountains of food wasted as coronavirus scrambles supply chain. *The Guardian.* Retrieved from https://www.theguardian.com/world/2020/apr/ 09/us-coronavirus-outbreak-agriculture-food-supply-waste

Canfield, D. E. (1998). A new model for Proterozoic ocean chemistry. *Nature, 396*(6710), 450–453.

Caporgno, M. P., & Mathys, A. (2018). Trends in microalgae incorporations into innovative food products with potential health benefits. *Frontiers in Nutrition, 5,* 58.

Carlson, J., & Chappell, J. (2015). *Deepening food democracy: The tools to create a sustainable, food secure and food sovereign future are already here—deep democratic approaches can show us how.* Minneapolis, MN: Institute for Agriculture and Trade.

Carson, R. (1962). *Silent spring.* Boston, MA: Houghton Mifflin.

Cerrado Manifesto. (2017). The future of the cerrado in the hands of the market: Deforestation and native vegetation conversion must be stopped. Retrieved from https:// d3nehc6yl9qzo4.cloudfront.net/downloads/ cerradomanifesto_september2017_atualizadooutubro.pdf

Charlton, K., Russell, J., Gorman, E., Hanich, Q., Deslisle, A., Campbell, B., & Bell, J. (2016). Fish, food security and health in Pacific Island countries and territories: A systematic literature review. *BMC Health*, *16*, 285.

Churchill, W. (1932). Fifty years hence. In *Thoughts and Adventures* (pp. 24–27). London: Thorton Butterworth.

Clapp, J. (2014). Financialization, distance, and global food politics. *Journal of Peasant Studies*, *41*, 5.

Clapp, J. (2016). *Food* (2nd ed.). Cambridge: Polity Press.

Clapp, J. (2020). Spoiled milk, rotten vegetables and a very broken food system. *New York Times*. Retrieved from https://www.nytimes.com/2020/05/08/opinion/coronavirus-global-food-supply.html#click=https://t.co/V7lrlyFeQx

Clarke, M. (2019). Pacific leaders, Australia agree to disagree about action on climate change. *ABC News*. Retrieved from https://www.abc.net.au/news/2019-08-15/no-endorsements-come-out-of-tuvalu-declaration/11419342

Clarkson, V. (2020). *Sugar in Australia: A food system approach: Competing issues, diverse voices, and rethinking pathways to a sustainable transition.* Retrieved from https://www.georgeinstitute.org.au/sugar-in-australia-a-food-systems-approach

Coggan, P. (2020). *More: The 10,000-year rise of the world economy.* London: Profile Books.

Compassion Over Killing. (2019). Tyson's "plant-based" news isn't plant-based at all. *Compassion Over Killing*. Retrieved from https://cok.net/tysons-plant-based-news-not-plant-based/

Coopes, A. (2019). Climate change, social determinants focus for WHO's new Healthier Populations pillar. *Croakey*. Retrieved from https://croakey.org/climate-change-social-determinants-focus-for-whos-new-healthier-populations-pillar/

Cota, G. M. (2016). *Disrupting maize: food, biotechnology, and nationalism in contemporary Mexico*. London: Rowman & Littlefield.

Couldry, N. (2010). *Why voice matters: Culture and politics after neoliberalism*. London: Sage.

Crane, R. (2018). Experts say algae is the food of the future. Here's why. *CNN Business*. Retrieved from https://money.cnn.com/2018/06/01/technology/algae-food/index.html

Crosby, A. W. (2003). *The Columbian exchange: Biological and cultural consequences of 1492*. Westport, CT: Praeger.

Daems, E. (2016). Tanzanian farmers are facing heavy prison sentences if they continue their traditional seed exchange. *Mondiaal Nieuws*. Retrieved from https://www.mo.be/en/analysis/tanzanian-farmers-are-facing-heavy-prison-sentences-if-they-continue-their-traditional-seed

Dalrymple, L., & Hilliard, G. (2020). *The ethical omnivore*. Sydney, NSW: Murdoch Books.

Davies, A. (2019). Tough nut to crack: The almond boom and its drain on the Murray-Darling. *The Guardian*. Retrieved from https://www.theguardian.com/australia-news/2019/may/26/tough-nut-to-crack-the-almond-boom-and-its-drain-on-the-murray-darling

Davis, M. (2020). The promise of an Australian homecoming: What would make an acknowledgement of country more welcome. *The Monthly*, 8–11. July.

Davison, D., & Kinsman, K. (2020). How I got radicalized around food. *Food & Wine*. Retrieved from https://www.foodandwine.com/fwpro/devita-davison-foodlab-detroit-kat-kinsman-interview.

Day, L. (2019). Australia's dairy farmers issue warning as mass exodus continues. *ABC News*. Retrieved from https://www.abc.net.au/news/2019-06-26/dairy-farmers-mass-exodus-from-the-industry/11215730

De Pieri, S. (2019). The basin and the kill. *Arena*. Retrieved from https://arena.org.au/the-basin-and-the-kill-by-stefano-de-pieri/

De Sousa Santos, B., Nunes, J. A., & Meneses, M. P. (2007). Introduction: Opening up the cannon of knowledge and recognition of difference. In B. De Sousa Santos (Ed.), *Another knowledge is possible: Beyond northern epistemologies* (Vol. 3, pp. xvix–lxxi). London: Verso.

De Sousa Santos, B., (Ed.). (2007). *Cognitive justice in a global world: Prudent knowledges for a decent life*. Washington, DC: Lexington Books.

Declaration of Nyéléni. (2007). Retrieved from https://nyeleni.org/spip.php?article290. Accessed on July 20, 2020.

DeLoughrey, E. M. (2019). *Allegories of the Anthropocene*. Durham, NC: Duke University Press.

Delucchi, M. A., & Jacobson, M. Z. (2011). Providing all global energy with wind, water, and solar power, Part II: Reliability, system and transmission costs, and policies. *Energy Policy*, 39, 1170–1190.

Devlin, H. (2015). Was 1610 the beginning of a new human epoch?. *The Guardian*. Retrieved from https://www.theguardian.com/science/2015/mar/11/was-1610-the-beginning-of-a-new-human-epoch-anthropocene

Dijkhorst, H. V. (2018). Cerrado deforestation disrupts water systems and poses business risks for soy producers. *Chain Reaction Research*. Retrieved from https://chainreaction research.com/report/cerrado-deforestation-disrupts-water-systems-poses-business-risks-for-soy-producers/

Dobbs, R., Sawers, C., Thompson, F., Manyika, J., Woetzel, J., Child, P., & Spatharou, A. (2014). *Overcoming obesity: An initial economic analysis*. Retrieved from https://www.mckinsey.com/~/media/McKinsey/Business%20Functions/Economic%20Studies%20TEMP/Our%20Insights/How%20the%20world%20could%20better%20fight%20obesity/MGI_Overcoming_obesity_Full_report.ashx

Doherty, B., Ensor, J., Heron, T., & Prado, P. (2019). Food systems resilience: Towards an interdisciplinary research agenda. *Emerald Open Research*, *1*, 4.

Du Toit, A. (2001). Ethical trading: A force for improvement, or corporate whitewash?. *Natural Resource Perspectives*, Retrieved from https://www.odi.org/sites/odi.org.uk/files/odi-assets/publications-opinion-files/2819.pdf

Duarte, C. M., Marbá, N., & Holmer, M. (2007). Rapid domestication of marine species. *Science*, *316*, 382–383.

Duarte, C. M., Holmer, M., Olsen, Y., Soto, D., Marbá, N., Guiu, J., & Karakassis, I. (2009). Will the oceans help feed humanity?. *BioScience*, *59*(11), 967–976.

Duhamel, K. (2018). Gakina Gidagwi'igoomin Anishinaabewiyang: We are all treaty people. *Canada's History*. Retrieved from https://www.canadashistory.ca/explore/settlement-immigration/gakina-gidagwi-igoomin-anishinaabewiyang-we-are-all-treaty-people

Dunford, R. (2017). Peasant activism and the rise of food sovereignty: Decolonising and democratising norm diffusion?. *European Journal of International Relations*, *23*(1), 145–167.

Eenennaam, A. L. V. (2019). Alternative meats and alternative statistics: What do the data say?. Paper presented at the Range Beef Cow Symposium XXVI, Mitchell, NE.

Eisen, M. (2018). How GMOs can save civilization (and probably already have). *Medium*. Retrieved from https://medium.com/impossible-foods/how-gmos-can-save-civilization-and-probably-already-have-6e6366cb893

Elver, H. (2017). Pesticides and the right to food. A/HRC/34/48. Office of the High Commissioner. United Nations Human Rights. Retrieved from https://hilalelver.org/resources/thematicreports/pesticides-and-food/

Escobar, A. (2017). *Designs for the pluriverse: Radical interdependence, autonomy, and the making of worlds*. Durham: Duke University Press.

Estes, N. (2019). *Our history is the future: Standing Rock versus the Dakota Access Pipeline, and the long tradition of Indigenous resistance*. London: Verso.

European Commission. (2017). *Food from the oceans: How can more food and biomass be obtained from the oceans in a way that does not deprive future generations of their benefits?*. Retrieved from https://ec.europa.eu/research/sam/pdf/sam_food-from-oceans_report.pdf

European Commission. (2018). Final report of the High Level Panel of the European decarbonisation pathways initiative. Retrieved from https://ec.europa.eu/info/publications/final-report-high-level-pnel-european-decarbonisation-pathways-initiative_en

Evans, A., & Miele, M. (2017). Food labelling as a response to political consumption: Effects and contradictions. In B. H. Margit Keller, T. A. Wilska, & M. Truninger (Eds.), *Routledge handbook on consumption* (pp. 233–247). London: Routledge.

Every one. Every day. (2020). Retrieved from https://www.weareeveryone.org/

Falzon, K. (2016). World's first farm to use solar power and seawater opens in Australia. *EcoWatch*. Retrieved from https://www.ecowatch.com/sundrop-farms-solar-desalination-2033987160.html

FAO. (2014). Principles on responsible agricultural investment that respects. rights livelihoods and resources. Retrieved from http://www.fao.org/3/a-au866e.pdf

FAO. (2015). Voluntary guidelines for securing sustainable small-scale fisheries. Retrieved from http://www.fao.org/voluntary-guidelines-small-scale-fisheries/en/

FAO. (2018). The state of world fisheries and aquaculture 2018-meeting the Sustainable Development Goals. Retrieved from http://www.fao.org/3/I9540EN/i9540en.pdf

FAO. (2019). Trophic levels. Retrieved from http://www.fao.org/fishery/topic/4210/en

FAO. (2020a). Mitigating risks to food systems during Covid-19: Reducing food loss and waste. Retrieved from http://www.fao.org/3/ca9056en/ca9056en.pdf

FAO. (2020b). The state of food security and nutrition in the world 2020: Transforming food systems for affordable diets. Retrieved from http://www.fao.org/documents/card/en/c/ca9692en

Farmer, P., Nizeye, B., Stulac, S., & Keshavjee, S. (2006). Structural violence and clinical medicine. *PLoS Med. 3*, e449.

Farmer, P. (2004). *Pathologies of power*. Berkeley, CA: University of California Press.

Fermanich, L. (2018). Decolonizing the food system. *New Entry Sustainable Farming Project*. Retrieved from https://nesfp.org/updates/2018/2/1/decolonizing-food-system

Fernandes, B. M. (2015). The formation and territorialisation of the MST in Brazil. In M. Carter (Ed.), *Challenging Social Inequality: The landless rural workers movement and agrarian reform in Brazil* (pp. 115–148). Durham, NC: Duke University Press.

Ferrante, L., & Fearnside, P. M. (2019). Brazil's new president and 'ruralists' threaten Amazonia's environment, traditional peoples and the global climate. *Environmental Conservation, 46,* 261–263.

FIAN International. (2010). *Why we oppose the principles for responsible agricultural investment (RAI).* Heidelberg: FIAN International.

FIAN International. (2020a). *Impact of Covid-19 on the human right to food and nutrition.* Retrieved from https://www.fian.org/files/files/Preliminary_monitoring_report_-_Impact_of_COVID19_on_the_HRtFN_EN(1).pdf.

FIAN International. (2020b). *SOFI report acknowledges urgent need for food systems transformation.* Retrieved from https://www.fian.org/en/press-release/article/sofi-report-acknowledges-urgent-need-for-food-systems-transformation-2531

Figueroa-Helland, L., Thomas, C., & Pérez Aguilera, A. (2018). Decolonizing food systems: Food sovereignty, Indigenous revitalization, and agroecology as counter-hegemonic movements. *Perspectives on Global Development & Technology, 17,* 173–201.

Fisher, A. (2020). Hunger incorporated: Who benefits from anti-hunger efforts?. In S. Jayaraman & K. De Master (Eds.), *Bite Back: People taking on corporate food and winning* (pp. 147–158). Oakland, CA: University of California Press.

Flanagan, T., Wilkie, M., & Iuliano, S. (2003). Australian South Sea Islanders: A century of race discrimination under Australian law. Retrieved from https://humanrights.gov.au/our-work/race-discrimination/publications/australian-south-sea-islanders-century-race

Flood, C., & Sloan, M. R. (Eds.). (2019). Food: Bigger than the plate. London: V&A Publishing.

Foer, J. S. (2019). We are the weather: Saving the planet begins at breakfast. London: Penguin.

Fortune, A. (2019a). Grant offered to lab-grown kangaroo meat start-up. Food Navigator-Asia. Retrieved from https://www.foodnavigator-asia.com/Article/2019/08/22/Grant-offered-to-lab-grown-kangaroo-meat-start-up

Fortune, A. (2019b). US plant-based market now worth $4.5bn. GlobalMeatNews.com. Retrieved from https://www.globalmeatnews.com/Article/2019/07/16/US-plant-based-market-now-worth-4.5bn/?utm_source=Newsletter_SponsoredSpecial&utm_medium=email&utm_campaign=Newsletter%2BSponsoredSpecial&c=16HhEPChu5CfTq HPeZn9oIXZvdgWpnP7

Foster, J. B., & Clark, B. (2020). The robbery of nature: Capitalism and the ecological rift. New York: Monthly Review Press.

Franklin-Wallis, O. (2019). White gold: The unstoppable rise of alternative milks. The Guardian. Retrieved from https://www.theguardian.com/news/2019/jan/29/white-gold-the-unstoppable-rise-of-alternative-milks-oat-soy-rice-coconut-plant

Fraser, E. (2020). Coronavirus: The perils of our 'just enough, just in time' food system. The Conversation. Retrieved from https://theconversation.com/coronavirus-the-perils-of-our-just-enough-just-in-time-food-system-133724

Frayssinet, F. (2015). The dilemma of soy. *Third World Resurgence*. Retrieved from https://www.twn.my/title2/resurgence/2015/295/cover05.htm

Frédéric, L., Hite, A. H., & Gregorini P. (2020). Livestock in evolving foodscapes and thoughtscapes. *Frontiers in Sustainable Food Systems*, *4*, Retrieved from https://www.frontiersin.org/article/10.3389/fsufs.2020.00105

Freire, P. (1970). *Pedagogy of the oppressed*. New York, NY: Bloomsbury.

Freshour, C. (2019). Cheap meat and cheap work in the U.S. Poultry industry: Race, gender, and immigration in corporate strategies to shape labour. In B. Winders, & E. Ransom (Eds.), *Global meat: Social and environmental consequences of the expanding meat industry* (pp. 121–140). Cambridge, MA: The MIT Press.

Friedmann, H., & McMichael, P. (1989). Agriculture and the state system: The rise and decline of national agricultures, 1870 to the present. *Sociologia Ruralis*, *29*(2), 93–117.

Friedrichs, J. (2013). *The future is not what it used to be: Climate change and energy scarcity*. Cambridge: MIT Press.

Friel, S. (2019). *Climate change and the people's health*. Oxford: Oxford University Press.

FSIN. (2019). Global report on food crises. Retrieved from https://www.fsinplatform.org/report/global-report-food-crisis-2019/

Fuller, B. H. (1969). *Operating manual for Spaceship Earth*. New York, NY: E.P. Dutton & Co.

Fulton, J., Norton, M., & Shilling, F. (2019). Water-indexed benefits and impacts of California almonds. *Ecological Indicators*, *96*(1), 711–717.

Future Food Institute. (2014). *Hackathon*. Retrieved from https://futurefood.network/institute/hackathon/

Galeano, E. (1971). *Las venas abiertas de América Latina*. Mexico: Siglo XXI Editores.

Gálvez, A. (2018). *Eating NAFTA: Trade, food policies, and the destruction of Mexico*. Oakland, CA: University of California Press.

Gammage, B. (2012). *The biggest estate on Earth: How aborigines made Australia*. Sydney, NSW: Allen & Unwin.

Garcia, D., Galaz, V., & Daume, S. (2019). EATLancet vs yes2meat: The digital backlash to the planetary health diet. *The Lancet*, *394*, 2153–2154.

Garg, S., Kim, L., Whitaker, M., O'Halloran, A., Cummings, C., Holstein, R., … Fry, A. (2020, March 1–30). Hospitalization rates and characteristics of patients hospitalized with laboratory-confirmed coronavirus disease 2019—COVID-NET, 14 States, 2020. *Morbidity and Mortality Weekly Report 2020*, *69*, 458–464.

Garnaut, R. (2019). *Super-power: Australia's low-carbon opportunity*. Carlton, VIC: Black Inc.

GBRMPA. (2019). Position statement: Climate change. Retrieved from http://elibrary.gbrmpa.gov.au/jspui/bitstream/11017/3460/5/v1-Climate-Change-Position-Statement-for-eLibrary.pdf

Gerber, P. J., Steinfeld, H., Henderson, B., Mottet, A., Opio, C., Dijkman, J., & Tempio, G. (2013). Tackling climate change through livestock – a global assessment of emissions and mitigation opportunities. Retrieved from http://www.fao.org/3/i3437e/i3437e.pdf

Ghosh, A. (2016). *The great derangement: Climate change and the unthinkable*. Chicago, IL: The University of Chicago Press.

Ghosh, I. (2020). Hunger Pandemic: The Covid-19 effect on global food insecurity. *Visual Capitalist*. Retrieved from https://www.visualcapitalist.com/covid-19-global-food-insecurity/

Gibbs, H. K., Rausch, L., Munger, J., Schelly, I., Morton, D. C., Noojipady, P., ... Walker, N. F. (2015). Brazil's Soy moratorium. *Science*, *347*(6220), 377–378.

Gibson-Graham, J. (1996). *The end of capitalism (as we know it): A feminist critique of political economy*. Minneapolis, MN: University of Minnesota Press.

Giggs, R. (2020). *Fathoms: The world in the whale*. Melbourne, VIC: Scribe.

Glickman, W. (2020). New York's rising tides: Climate inequality and Sandy's legacy. Retrieved from https://www.nybooks.com/daily/2020/06/13/new-yorks-rising-tides-climate-inequality-and-sandys-legacy/?utm_medium=email&utm_campaign=NYR%20Unpresident&utm_content=NYR%20Unpresident+CID_48ee47bbdce091b9ebee500827e60701&utm_source=Newsletter&utm_term=New%20York

Global Footprint Network. (2020). *Earth overshoot day*. Retrieved from https://www.overshootday.org/2020-calculation/

Gómez-Barris, M. (2017). *The extractive zone: Social ecologies and decolonial perspectives*. Durham: Duke University Press.

Goodall, J. R. (2019). *The politics of the common good*. Sydney, NSW: NewSouth Publishing.

Goodell, J. (2019). The world's most insane energy project moves ahead. *Rolling Stone*. Retrieved from https://www.rollingstone.com/politics/politics-news/adani-mine-australia-climate-change-848315/

Goodyear, D. (2015). What milk should I drink?. *The New Yorker*. October 23. Retrieved from https://www.newyorker.com/news/daily-comment/what-milk-should-i-drink

Gordon, C., & Hunt, K. (2018). Reform, justice, and sovereignty: A food systems agenda for environmental communication. *Environmental Communication.*, *13*(1), 9–22.

GRAIN. (2010). Responsible farmland investing? Current efforts to regulate land grabs will make things worse. *Against the Grain*. Retrieved from https://www.grain.org/article/entries/4564-responsible-farmland-investing-current-efforts-to-regulate-land-grabs-will-make-things-worse

GRAIN. (2015). Foreign pension funds and land grabbing in Brazil. *Against the Grain*. Retrieved from https://www.grain.org/article/entries/5336-foreign-pension-funds-and-land-grabbing-in-brazil

GRAIN. (2019). Step aside agribusiness: It's time for real solutions to the climate crisis. Retrieved from https://www.localfutures.org/step-aside-agribusiness-its-time-for-real-solutions-to-the-climate-crisis/

Grandin, T. (2020). Big meat supply chains are fragile. *Forbes*. Retrieved from https://www.forbes.com/sites/templegrandin/2020/05/03/temple-grandin-big-meat-supply-chains-are-fragile/amp/?fbclid=IwAR1GCXzuoJd7qq7AdjlzXjU2-5FsU_HBU6UoDRyaFnSRNsS1loPaV3Xk8xg

Grauer, Y. (2010). Plant diversity necessary. *Bee Culture*, *138*(7), 37.

Green, L. (2020). Oceanic regime shift. In E. Probyn, K. Johnston, & N. Lee (Eds.), *Sustaining seas: Oceanic space and the politics of care* (pp. 11–26). London: Rowman & Littlefield.

Greenberg, P. (2010). *Four fish: The future of the last wild food*. London: Penguin.

Greenebaum, J. (2018). Vegans of color: Managing visible and invisible stigmas. *Food, Culture & Society, 21*(5), 680–697.

Greenwood, A. (2018). 'Eat less meat' ignores the role of animals in the ecosystem. *Civil Eats*. Retrieved from https://civileats.com/2018/01/26/eat-less-meat-ignores-the-role-of-animals-in-the-ecosystem/

Guha, R., & Martínez-Alier, J. (1997). *Varieties of environmentalism: Essays North and South*. London: Earthscan.

Guthman, J. (2008). Bringing good food to others: Investigating the subjects of alternative food practice. *Cultural Geographies, 15*(4), 431–447.

Guthman, J. (2011). *Weighing in: Obesity, food justice and the limits of capitalism*. Berkeley, CA: University of California Press.

Gwynn, K. S., Searle, T., Senior, A., Lee, A., & Brimblecombe, J. (2019). Effect of nutrition interventions on diet-related and health outcomes of Aboriginal and Torres Strait Islander Australians: A systematic review. *BMJ Open, 9*.

Hannam, P. (2019). 'On their knees': Drought and nuts blamed for decimating food sector. *The Sydney Morning Herald*. Retrieved from https://www.smh.com.au/national/on-their-knees-drought-and-nuts-blamed-for-decimating-food-sector-20191024-p53441.html

Hannesson, R. (2008). *Privatization of the oceans: Handbook of marine fisheries conservation and management*. New York, NY: Oxford University Press.

Harari, Y. N. (2019). *Sapiens*. Melbourne, VIC: Penguin.

Haraway, D. (2015). Anthropocene, Capitalocene, Plantationocene, Chthulucene: Making kin. *Environmental Humanities, 6*, 159–165. doi:10.1215/22011919-3615934

Haraway, D. (2019). Symbiogenesis, sympoiesis, and art science activisms for staying with the trouble. In A. Tsing, H. Swanson, E. Gan, & N. Bubandt (Eds.), *Arts of living on a damaged planet* (pp. M25–M50). Minneapolis, MN: University of Minnesota Press.

Harris, E. (2009). Neoliberal subjectivities or a politics of the possible? Reading for difference in alternative food networks. *Area, 41*(1), 55–63.

Harrison, J. L. (2011). *Pesticide drift and the pursuit of environmental justice*. Cambridge, MA: Massachusetts Institute of Technology.

Harvey, D. (2015). *Seventeen contradictions and the end of capitalism*. Croydon: Profile Books.

Hawken, P. (2007). *Blessed unrest: How the largest social movement in history is restoring grace, justice, and beauty to the world*. New York, NY: Penguin.

Hawkes, S. F., Lobstein, T., & Lang, T. (2012). Linking agricultural policies with obesity and noncommunicable diseases: A new perspective for a globalising world. *Food Policy, 37*(3), 343–353.

Hawkes, C. (2006). Uneven dietary development: Linking the policies and processes of globalization with the nutrition transition, obesity and diet-related chronic diseases. *Globalization and Health, 2*, 4.

Hayman, E. (2018). Future rivers of the Anthropocene or whose Anthropocene is it? Decolonising the Anthropocene. *Decolonization: Indigeneity, Education & Society*, 6(2), 77–92.

Herring, R., & Paarlberg, R. (2016). The political economy of biotechnology. *Annual Review of Resource Economics*, 8(1), 397–416.

Hickel, J. (2020). The racist double standards of international development. *Al Jazeera.com*. Retrieved from https://www.aljazeera.com/indepth/opinion/racist-double-standards-international-development-200707082924882.html

Hightower, J. (1975). *Eat your heart out*. New York, NY: Vintage.

Hilary, J. (2013). *The poverty of capitalism: Economic meltdown and the struggle for what comes next*. London: Pluto Press.

Hill, R., Grant, C., George, M., Robinson, C. J., Jackson, S., & Abel, N. (2012). Typology of Indigenous engagement in Australian environmental management: Implications for knowledge integration and social-ecological system sustainability. *Ecology & Society*, 17(1), 23.

Hird, M. J., & Zahara, A. (2017). The Arctic wastes. In R. Gruisin (Ed.), *Anthropocene feminism* (pp. 121–146). Minneapolis, MN: University of Minnesota.

Hocquette, J. F. (2016). Is in vitro meat the solution for the future?. *Meat Science*, 120, 167–176. doi:10.1016/j.meatsci.2016.04.036

Holmes, S. (2020). Farmworkers are dying, COVID-19 cases are spiking, and the food system is in peril. *Salon*. Retrieved from https://www.salon.com/2020/05/31/farmworkers-are-dying-covid-19-cases-are-spiking-and-the-food-system-is-in-peril/

Holt-Giménez, E. (2002). Measuring farmers' agroecological resistance after Hurricane Mitch in Nicaragua: A case study in participatory, sustainable land management impact monitoring. *Agriculture, Ecosystems and Environment, 93*, 87–105.

Holt-Giménez, E. (2017). *A foodie's guide to capitalism: Understanding the political economy of what we eat.* New York, NY: Monthly Review Press.

Honeybone, E., & McAllister, K. (2019). Adopting an Indigenous totem could be a simple way to care for country. *ABC News.* Retrieved from https://www.abc.net.au/news/2019-06-01/every-aussie-should-adopt-a-totem/11116978

Hoover, E. (2017). *The river is in Us: Fighting toxics in a Mohawk community.* Minneapolis, MN: University of Minnesota Press.

Howard, P. H. (2019). Corporate concentration in global meat processing: The role of feed and finance subsidies. In B. Winders, & E. Ransom (Eds.), *Global meat: Social and environmental consequences of the expanding meat industry* (pp. 31–54). Cambridge, MA: The MIT Press.

Hughes, N. (2019). Drought and climate change are driving high water prices in the Murray-Darling Basin. *ABC News.* Retrieved from https://www.abc.net.au/news/2019-07-22/drought-and-climate-change-are-driving-high-water-prices-in-the/11329502

Hughes, L. (2020). The milk of human genius: On the end of the cow and the future of food. *The Monthly, March*, (pp. 44–50).

IDMC. (2019). Global report on internal displacement 2019. Retrieved from https://www.internal-displacement.org/global-report/grid2019/

IEA. (2020). Global energy demand to plunge this year as a result of the biggest shock since the Second World War. Retrieved from https://www.iea.org/news/global-energy-demand-to-plunge-this-year-as-a-result-of-the-biggest-shock-since-the-second-world-war

Ifansasti, U. (2012). Death metal: Tin mining in Indonesia - in pictures. *The Guardian*. Retrieved from https://www.theguardian.com/environment/gallery/2012/nov/23/tin-mining-indonesia-bangka-photographs

Impossible Foods. (2019). FAQ: Does it contain genetically modified ingredients?. Retrieved from https://faq.impossiblefoods.com/hc/en-us/articles/360023038894-Does-the-Impossible-Burger-contain-genetically-modified-ingredients-

IPCC. (2019). Climate Change and Land: An IPCC Special Report on climate change, desertification, land degradation, sustainable land management, food security, and greenhouse gas fluxes in terrestrial ecosystems. Retrieved from https://www.ipcc.ch/site/assets/uploads/2019/08/4.-SPM_Approved_Microsite_FINAL.pdf

IPES. (2017). Too big to feed: Exploring the impacts of mega-mergers, consolidation and concentration of power in the agri-food sector. Retrieved from www.ipes-food.org/images/Reports/Concentration_FullReport.pdf

IPES-Food. (2020). COVID-19 and the crisis in food systems: Symptoms, causes, and potential solutions. Retrieved from http://www.ipes-food.org/_img/upload/files/COVID-19_CommuniqueEN.pdf

Itzkan, S. (2020). Opinion: Software to swallow – impossible foods should be called impossible patents. *The Medium*. Retrieved from https://medium.com/@sethitzkan/opinion-software-to-swallow-impossible-foods-should-be-called-impossible-patents-71805ecec9de

Jahren, H. (2020). *The story of more: How we got to climate change and where to go from here*. London: Fleet.

Jayaraman, S. & De Master, K. (2020) *Bite Back: People taking on corporate food and winning.* Oakland, CA: University of California Press.

Jayaraman, S. (2020). Food workers taking on Goliath. In S. Jayaraman & K. De Master (Eds.), *Bite back: People taking on corporate food and winning* (pp. 107–118). Oakland, CA: University of California Press.

Johnson, J., Bell, J., & De Young, C. (2013). Priority adaptations to climate change for Pacific fisheries and aquaculture: Reducing risks and capitalizing on opportunities. FAO/Secretariat of the Pacific Community Workshop, 5–8 June, 2012, Noumea, New Caledonia. FAO Fisheries and Aquaculture Proceedings No. 28. Rome: FAO.

Johnson, H. (2007). *Two families: Treaties and government.* Saskatoon, SK: Purich.

Kallis, G. (2011). In defence of degrowth. *Ecological Economics, 70,* 873–880.

Kammal, A. G., Linklater, R., Thompson, S., Dipple, J., & Ithinto Mechisowin Committee. (2015). A recipe for change: Reclamation of Indigenous food sovereignty in *O-Pipon-Na-Piwin Cree* Nation for decolonization, resource sharing, and cultural restoration. *Globalizations, 12*(4), 559–575.

Kateman, B. (2019). Yes, algae is green and slimy - but it could also be the future of food. *The Guardian.* Retrieved from https://www.theguardian.com/commentisfree/2019/aug/13/algae-spirulina-e3-blue-majik-health-benefits

Kaufman, A. C. (2018). The warped environmentalism of America's biggest industrial meat producer. *Huffpost.* Retrieved from https://www.huffingtonpost.com.au/entry/tyson-foods-environmentalism-regulation-sustainability_n_5a9562d6e4b0699553cc7656?ri18n=true

Kennedy, E. H., Johnston, J., & Parkins, J. R. (2018). "Small-p politics: How pleasurable, convivial and pragmatic political ideals influence engagement in eat-local initiatives. *British Journal of Sociology*, *69*(3), 670–690.

Kim, T., Yong, H., Kim, Y., Kim, H., & Choi, Y. (2019). Edible insects as a protein source: A review of public perception, processing technology, and research trends. *Food Science of Animal Resources*, *39*(4), 521–540.

Kite-Powell, J. (2018). See how algae could change our world. *Forbes*. Retrieved from https://www.forbes.com/sites/jenniferhicks/2018/06/15/see-how-algae-could-change-our-world/#6e4320cb3e46

Klein, N. (2014). *This changes everything: Capitalism vs the climate*. New York, NY: Simon & Schuster.

Klinenberg, E. (2018). *Palaces for People: How to build a more equal and united society*. London: Penguin.

Kolbert, E. (2014). *The sixth extinction: An unnatural history*. London: Bloomsbury.

Krimsky, S. (2019). *GMOs decoded: A skeptics view of genetically modified foods*. Cambridge, MA: The MIT Press.

Kuhn, K., & Andersen, K. (2015). *The sustainability secret: Our diet to transform the world*. San Rafael, CA: Earth Aware Editions.

Kurlansky, M. (1998). *Cod: The biography of the fish that changed the world*. Sydney, NSW: Penguin Random House.

La Vía Campesina. (2017). Nyéléni newsletter: Oceans and water. No 31. Retrieved from https://nyeleni.org/spip.php?rubrique186

La Vía Campesina. (2020). Amid COVID-19, Vía Campesina says stay home but not silent. Retrieved from https://grassrootsonline.org/in-the-news/amid-covid-19-via-campesina-says-stay-home-but-not-silent/

Lang, T. (2020). *Feeding Britain: Our food problems and how to fix them*. London: Pelican.

Langert, B. (2019). *The battle to do good: Inside McDonald's sustainability journey*. Bingley: Emerald Publishing.

Lappé, F. M. (1971). *Diet for a small planet*. New York, NY: Ballantine Books.

Laroche, P. C. J. S., Schulp, C. J. E., Kastner, T., & Verburg, P. H. (2020). Telecoupled environmental impact of current and alternative Western diets. *Global Environmental Change*, 62, 102066.

Latour, B. (2019). *Down to Earth: Politics in the new climatic regime*. Cambridge: Polity Press.

Le Feon, V., Burel, F., Chifflet, R., Henry, M., Richroch, A., Vaissiere, B., & Baudry, J. (2010). Solitary bee abundance and species richness in dynamic agricultural landscapes. *Agriculture, Ecosystems and Environment*, 166, 94–101.

Le Guin, U. K. (2019). Deep in admiration. In A. Tsing, H. Swanson, E. Gan, & N. Bubandt (Eds.), *Arts of living in a damaged planet* (pp. M15–M21). MN: University of Minnesota Press.

Lengnick, L. (2015). *Resilient agriculture: Cultivating food systems for a changing climate*. Gabriola Island: New Society Publishers.

Leroy, F., & Cohen, M. (2019). The EAT-Lancet Commission's controversial campaign: A global powerful action against meat?. *European Food Agency News*. Retrieved from https://Q6 www.efanews.eu/item/6053

Lichtenstein, N. (2005). Wal-Mart: Template for 21st century capitalism?. *New Labour Forum, 14*(1), 21–30.

Lima, M., Da Silva, C. A., Rausch, L., Gibbs, H. K., & Johann, J. A. (2019). Demystifying sustainable soy in Brazil. *Land Use Policy, 82*, 349–352.

Little Bear, L. (2000). Jagged worldviews colliding. In M. Battiste (Ed.), *Reclaiming Indigenous voice and vision* (pp. 77–85). Vancouver, BC: University of British Columbia Press.

Lock the Gate. (n.d). Lock the Gate Alliance. Retrieved from https://www.lockthegate.org.au/about_us

Lonetree, A. (2012). *Decolonizing museums: Representing native America in national and tribal museums.* Chapel Hill, NC: University of North Carolina Press.

Lovelock, J., & Margulis, L. (1974). Atmospheric homeostasis by and for the biosphere: The Gaia Hypothesis. *Tellus, 26*, 1–2.

Lucas, A. N. A. (2019). Beyond Meat uses climate change to market fake meat substitutes. *Scientists are Cautious.* Retrieved from https://www.cnbc.com/2019/09/02/beyond-meat-uses-climate-change-to-market-fake-meat-substitutes-scientists-are-cautious.html

Lynch, J., & Pierrehumbert, R. (2019). Climate impacts of cultured meat and beef cattle. *Frontiers in Sustainable Food Systems, 3*, 5. doi:10.3389/fsufs.2019.00005

Lyster, R. (1974). PrimeMinister Scott Morrison and the Pacific Island Forum. *SEI Magazine: Voicing Community, 3*, 16–20.

Maddison, S. (2019). *The colonial fantasy: Why white Australia can't save Black problems.* Sydney, NSW: Allen and Unwin.

Malm, A. (2016). *Fossil capital: The rise of steam power and the roots of global warming.* London: Verso.

Mann, M., & Toles, T. (2016). *The Madhouse effect: How climate change denial is threatening our planet, destroying our politics, and driving us crazy*. New York, NY: Columbia University Press.

Mann, A. (2014). *Global activism in food politics: Power shift*. Basingstoke: Palgrave Macmillan.

Mann, A. (2017). Food sovereignty and the politics of food scarcity. In M. C. Dawson, C. Rosin, & N. Wald (Eds.), *Global resources scarcity: Catalyst for conflict or cooperation?* (pp. 131–145). London: Routledge.

Mann, A. (2019). *Voice and participation in global food politics*. Abingdon: Routledge.

Mann, A. (2020a). The protection of small-scale fisheries in global policy-making through food sovereignty. In E. Probyn, K. Johnston, & N. Lee (Eds.), *Sustaining seas: Oceanic space and the politics of care* (pp. 185–200). Lanham, MD: Rowman & Littlefield.

Mann, A. (2020b). Are you local? Digital inclusion in participatory foodscapes. In D. Lupton & Z. Feldman (Eds.), *Digital food cultures* (pp. 147–161). London: Routledge.

Mann, A. (2020c). Hacking the foodscape: Digital communication in the co-design of sustainable and inclusive food environments. In J. Diaz-Pont, P. Maeseele, A. Egan Sjolander, M. Mishra, & K. Foxwel-Norton (Eds.), *The local and the digital in environmental communication* (pp. 183–202). Cham: Palgrave Macmillan.

Markard, J., Raven, R., & Truffer, B. (2012). Sustainability transitions: An emerging field of research and its prospects. *Research Policy, 41*, 955–967.

Marsham, J. (2020). East Africa faces triple crisis of Covid-19, locusts and floods. *Climate Change News*. Retrieved from https://www.climatechangenews.com/2020/05/11/east-africa-faces-triple-crisis-covid-19-locusts-floods/

Martin, L. (2019). Scott Morrison to sell Pacific 'step up' on Solomons visit as pressure builds over climate. *The Guardian.com*. Retrieved from https://www.theguardian.com/australia-news/2019/may/27/scott-morrison-expected-to-get-a-whack-on-climate-change-on-solomons-visit

Martínez-Alier, J. (2003). *The environmentalism of the poor: A study of ecological conflicts and valuation*. Cheltenham: Edward Elgar.

Martínez-Torres, M., & Rosset, P. (2014). Diàlogo de saberes in La Via Campesina: Food sovereignty and agroecology. *The Journal of Peasant Studies, 41*(6), 979–997.

Marx, K. (1976). *Capital* (Vol. 1). New York, NY: Vintage.

Massy, C. (2017). *Call of the reed warbler: A new agriculture, a new Earth*. St Lucia, QLD: University of Queensland Press.

Mayes, C. (2018). *Unsettling food politics: Agriculture, dispossession and sovereignty in Australia*. London: Rowman & Littlefield.

McDonald, J. (2019). Mergers in seeds and agricultural chemicals: What happened?. *Amber Waves*. Retrieved from https://www.ers.usda.gov/amber-waves/2019/february/mergers-in-seeds-and-agricultural-chemicals-what-happened/

McDonell, E. (2018). The quinoa boom goes bust in the Andes. Retrieved from https://nacla.org/news/2018/03/12/quinoa-boom-goes-bust-andes

McFall-Ngai, M. (2017). Noticing microbial worlds: The postmodern synthesis in biology. In H. S. Anna Tsing, E. Gan, & N. Bubandt (Eds.), *Arts of living on a damaged planet* (pp. M51–M69). Minneapolis, MN: University of Minnesota Press.

McGivney, A. (2020). 'Like sending bees to war': The deadly truth behind your almond milk obsession. *The Guardian*. Retrieved from https://www.theguardian.com/environment/2020/jan/07/honeybees-deaths-almonds-hives-aoe

McGuirk, J. (2014). *Radical cities: Across Latin America in search of a new architecture*. London: Penguin.

McKenna, M. (2017). *Big chicken: The incredible story of how antibiotics created modern agriculture and changed the way the world eats*. Washington, DC: National Geographic Partners LLC.

McKibben, B. (2019). *Falter*. Carlton: Black Inc.

McMichael, P. (2005). Global development and the corporate food regime. In F. H. Buttel, & P. McMichael (Eds.), *New directions in the sociology of global development* (Vol. 11, pp. 269–303). Bingley: Emerald Group Publishing Limited.

McMichael, P. (2009). A food regime genealogy. *The Journal of Peasant Studies*, 36(1), 139–169.

McMichael, P. (2010). *Contesting development*. New York, NY: Routledge.

McMichael, P. (2013). *Food regimes and agrarian questions*. Black Point, NS: Fernwood Publishing.

McMichael, P. (2020). The globalization project in crisis. *Alternate Routes*, 31(1). Retrieved from http://www.alternateroutes.ca/index.php/ar/article/view/22507

McMillan, T. (2012). *The American way of eating: Undercover at Walmart, Applebee's, farm fields and the dinner table.* New York, NY: Scribner.

Meadows, D. H., Meadows, D. L., Randers, J., & Behrens, W. W. (1972). *The limits to growth: A report for the club of Rome's project on the predicament of mankind.* London: Earth Island.

Meadows, D., Randers, J., & Meadows, D. (2004). *A synopsis: Limits to growth: The 30-year update.* Retrieved from http://donellameadows.org/archives/a-synopsis-limits-to-growth-the-30-year-update/

Medetshy, A., & Durisin, M. (2020). Exports of Russian wheat dry up, stoking food security concerns. *Aljazeera.* Retrieved from https://www.aljazeera.com/ajimpact/exports-russian-wheat-dry-stoking-food-security-concerns-200426172340195.html

Méndez, V. E., Bacon, C. M., & Cohen, R. (2013). Agroecology as a transdisciplinary, participatory, and action-oriented approach. *Agroecology and Sustainable Food Systems, 37*(1), 3–18.

Menser, M. (2014). The territory of self-determination, agroecological production, and the role of the state. In P. Andreé, J. Ayres, M. Bosia, & J. Massicotte (Eds.), *Globalization and food sovereignty* (pp. 53–83). Toronto, ON: University of Toronto Press.

Metelerkamp, L. (2020). US Agribusiness takes aim at global food policy reform. *Food Tank.* Retrieved from https://www.theguardian.com/environment/2020/apr/17/chicken-factory-tyson-arkansas-food-workers-coronavirus

Mexican Constitution. (1917). *Mexican Constitution of 1917.* The Library of Congress. Retrieved from https://www.loc.gov/item/17021628/

Meyer, G., & Schipani, A. (2019). Brazil set to overtake US as world's largest soyabean producer. *Financial Times.* Retrieved from https://www.ft.com/content/8b2bb828-1ad0-11ea-97df-cc63de1d73f4

Michail, N. (2019). Oatly developers ready for repeat success with patented quinoa milk: 'We can do it again with Quiny'. *FoodNavigator-USA.* Retrieved from https://www.foodnavigator.com/Article/2019/11/11/Oatly-developers-to-launch-patented-quinoa-milk?utm_source=newsletter_daily&utm_medium=email&utm_campaign=11-Nov-2019&c=16HhEPChu5Cr2V8I5nJz5aTnCgWw23AN&p2=

Mighty Earth. (2017). Mystery meat II: The industry behind the quiet destruction of the American Heartland. Retrieved from http://www.mightyearth.org/wp-content/uploads/2017/08/Meat-Pollution-in-America.pdf

Mills, E. (2018). Implicating 'fisheries justice' movements in food and climate politics. *Third World Quarterly, 39*(7), 1270–1289. doi:10.1080/01436597.2017.1416288

Mintz, S. (1986). *Sweetness and power: The place of sugar in modern history.* London: Penguin.

Mohorčich, J., & Reese, J. (2019). Cell-cultured meat: Lessons from GMO adoption and resistance. *Appetite, 143*, 104408.

Moore, J. W. (2015). *Capitalism in the web of life: Ecology and the accumulation of capital.* London: Verso.

Moore, J. W. (2017). The Capitalocene, Part I: On the nature and origins of our ecological crisis. *The Journal of Peasant Studies, 44*(3), 594–630. doi:10.1080/03066150.2016.1235036

Morena, M. A. (2020). Is veganism the solution to climate change? In *Right to food and nutrition watch. Overcoming ecological crises: Reconnecting food, nature and human rights* (pp. 46–57). Heidelberg: FIAN International & Brot für die Welt.

Morton, T. (2010). *The ecological thought*. Cambridge, MA: Harvard University Press.

Morton, T. (2013). *Hyperobjects: Philosophy and ecology after the end of the world*. Minneapolis, MN: University of Minnesota Press.

Mousseau, F., Currier, A., Fraser, E., & Green, J. (2016). *Driving dispossession: The global push to "unlock the economic potential of land"*. Oakland, CA: The Oakland Institute.

Mueller, M. L. (2017). *Being salmon, being human: Encountering the wild in us and us in the wild*. White River Junction, VT: Chelsea Green Publishing.

MUFPP. (2015). Milan Urban Food Policy Pact. Retrieved from http://www.milanurbanfoodpolicypact.org/viewed. Accessed on February 21, 2018.

Nargi, L. (2020). Community food co-ops are thriving during the pandemic. *Civil Eats*. Retrieved from https://civileats.com/2020/05/15/community-food-co-ops-are-thriving-during-the-pandemic/amp/?__twitter_impression=true

Nestle, M. (2007). *Food politics: How the food industry influences nutrition and health*. Berkeley, CA: University of California Press.

Newman, L. (2019). *Lost feast: Culinary extinction and the future of food*. Toronto, ON: ECW Press.

Newman, C. (2020). Food and empire. *The Medium*. Retrieved from https://medium.com/sylvanaquafarms/food-and-empire-844506392422

NFU. (2019). *Tackling the farm crisis and the climate crisis: A transformative strategy for Canadian farms and food systems.* Washington, DC: National Farmers Union.

Ng, M., Fleming, T., Robinson, M., Thomson, B., Graetz, N., Margono, C., ... Gakidou, E. (2014). Global, regional and national prevalence of overweight and obesity in children and adults during 1980-2013: A systematic analysis for the Global Burden of Disease Study 2013. *The Lancet, 384*(9945), 766–781. Retrieved from https://www.ncbi.nlm.nih.gov/pmc/articles/PMC4624264/.

Nixon, R. (2013). *Slow violence and the environmentalism of the poor.* Boston, MA: Harvard University Press.

Noble, D. (2014). Bee health is crucial for pollination. *Growing, 12*(4), 12.

Nolan, J., & Boersma, M. (2019). *Addressing modern slavery.* Sydney: NewSouth Publishing.

Nossiter, J. (2019). *Cultural insurrection: A manifesto for the arts, agriculture, and natural wine.* New York, NY: Other Press.

O'Malley, N., & Foley, M. (2020). Labor's climate policy: Olive branch or white feather?. *The Sydney Morning Herald.* Retrieved from https://amp.smh.com.au/politics/federal/labors-climate-policy-olive-branch-or-white-feather-20200626-p556if.html?__twitter_impression=true

Olsen, Y. (2015). How can mariculture better help feed humanity?. *Frontiers in Marine Science, 2*, 46.

Ord, T. (2020). *The precipice: Existential risk and the future of humanity.* London: Bloomsbury Publishing.

Oxfam. (2017). An economy for the 99%. Retrieved from https://www-cdn.oxfam.org/s3fs-public/file_attachments/bp-economy-for-99-percent-160117-en.pdf

Pachirat, T. (2013). *Every twelve seconds: Industrialised slaughter and the politics of sight*. New Haven, CT: Yale University Press.

Panayi, P. (2014). *Fish and chips: A history*. London: Reaktion Books Ltd.

Parenti, C. (2011). *Tropic of chaos: Climate change and the new geography of violence*. New York, NY: Bold Type Books.

Parker, C., & Johnson, H. (2019). From food chains to food webs: Regulating capitalist production and consumption in the food system. *Annual Review of Law Social Science*, *15*, 205–225.

Pascoe, B. (2014). *Dark Emu Black Seeds: Agriculture or accident?*. Broome, WA: Magabala Books Aboriginal Corporation.

Pascoe, B. (2018). Let's talk about real Australian food. Retrieved from https://www.sbs.com.au/food/article/2018/05/29/comment-lets-talk-about-real-australian-food

Patel, R., & de Wit, M. M. (2020). Call to action – the corporate stock in trade. In S. Jayaraman & K. De Master (Eds.), *Bite back: People taking on corporate food and winning* (pp. 175–191). Oakland, CA: University of California Press.

Patel, R., & Moore, J. (2018). *A history of the world in seven cheap things: A guide to capitalism, nature, and the future of the planet*. Melbourne, VIC: Black Inc.

Patel, R., Jayaraman, S., Boyd, J. W., Jr, Shute, L., Perls, D., & Carpenter, Z. (2017). The future of food. *The Nation*. Retrieved from https://www.thenation.com/article/archive/the-future-of-food/. Accessed on October 11, 2017.

Patel, R. (2009). *Stuffed and Starved: Markets, power and the hidden battle for the world food system*. Melbourne, VIC: Black Inc.

Pausz, T. (2015). Substitutions: Essay on food systems and materials science. Retrieved from https://www.pausz.org/Substitutions

PCFS. (2013). An initial statement on the Zero Draft of the Principles for Responsible Agricultural Investments (RAI) in the context of food security and nutrition. Retrieved from https://www.farmlandgrab.org/uploads/attachment/PCFS%20Critique%20on%20rai.pdf

Peet, R., Robbins, P., & Watts, M. (Eds.). (2011). *Global political ecology* (1st ed.). Abingdon, OX: Routledge.

Petrini, C. (2007). *Slow food nation: Why our food should be good, clean, and fair*. New York, NY: Rizzoli Ex Libris.

Philpott, T. (2020). *Perilous bounty: The looming collapse of American farming and how we can prevent it*. New York: Bloomsbury Publishing.

Piper, K. (2019). The unlikely partnership that might decide the future of meat: The country's biggest meat companies are investing in vegan foods. Here's why. Retrieved from https://www.vox.com/future-perfect/2019/3/22/18273892/tysonvegan-vegetarian-lab-meat-climate-change-animals

Polish, J. (2016). Decolonizing veganism: On resisting vegan whiteness and racism. In J. Castricano & R. R. Simonsen (Eds.), *Critical perspectives on veganism*. Cham: Palgrave.

Pollan, M. (2006). *The Omnivore's Dilemma: A natural history of four meals*. New York, NY: Penguin.

Pollan, M. (2020). The sickness in our food supply. *The New York Review of Books*. Retrieved from https://www.nybooks.com/articles/2020/06/11/covid-19-sickness-food-supply/

Pomeranz, K. (2000). *The great divergence: China, Europe, and the making of the modern world economy*. Princeton, NJ: Princeton University Press.

Poore, J., & Nemecek, T. (2018). Reducing food's environmental impacts through producers and consumers. *Science, 360*(6392), 987–992.

Povitz, L. D. (2019). *Stirrings: How activist New Yorkers ignited a movement for food justice*. Chapel Hill, NC: University of North Carolina Press.

Power, E., Black, J., & Brady, J. (2020). More than food banks are needed to feed the hungry during the coronavirus pandemic. *The Conversation*. Retrieved from https://theconversation.com/more-than-food-banks-are-needed-to-feed-the-hungry-during-the-coronavirus-pandemic-136164

Prager, A., & Milhorance, F. (2018). Cerrado: Agribusiness may be killing Brazil's 'birthplace of waters'. *Mongaby*. Retrieved from https://news.mongabay.com/2018/03/cerrado-agribusiness-may-be-killing-brazils-birthplace-of-waters/

Price, C. (1979, April 4, 2020). *Technology is the answer but what was the question?*. Retrieved from https://www.pidgeondigital.com/talks/technology-is-the-answer-but-what-was-the-question-/

Pringle, P. (2003). *Food Inc.: Mendel to Monsanto - the promises and perils of the biotech harvest*. New York, NY: Simon & Schuster.

Probyn, E. (2016). *Eating the ocean*. Durham, NC: Duke University Press.

Probyn, E. (2020). "The sea is empty": Fishers, migrants, and a watery humanism. In E. Probyn, K. Johnston, & N. Lee

(Eds.), *Sustaining the seas: Oceanic space and the politics of care*. London: Rowman & Littlefield.

Provenza, F. (2018). *Nourishment: What animals can teach us about rediscovering our nutritional wisdom*. London: Chelsea Green Publishing.

Queensland Government. (2010). *CoalPlan 2030: Laying the foundations of a future*. Retrieved from http://www.dlgrma.qld.gov.au/resources/plan/cg/coal-plan-2030.pdf

Raasch, S. (2017). Biosecurity vulnerabilities in crop monocultures. Propagate. Retrieved from https://www.propagate.org/research/2017/9/4/biosecurity-vulnerabilities-in-crop-monocultures

Ramsden, J. (2019). Same same, but different: Creating positive futures for Australian animal agriculture. *Farm Policy Journal, 16*, 4–11.

Ranganathan, J., Waite, R., Searchinger, T., & Hanson, C. (2018). How to sustainably feed 10 billion people by 2050, in 21 charts. *World Resources Institute*. Retrieved from https://www.wri.org/blog/2018/12/how-sustainably-feed-10-billion-people-2050-21-charts

Raphelson, S. (2019). Nobody is moving our cheese: American surplus reaches record high. *NPR*. Retrieved from https://www.npr.org/2019/01/09/683339929/nobody-is-moving-our-cheese-american-surplus-reaches-record-high

Raubenheimer, D., & Simpson, S. J. (2020). *Eat like the animals: What nature teaches us about the science of healthy eating*. Sydney, NSW: HarperCollins.

Raworth, K. (2018). *Donut economics: Seven ways to think like a 21st century economist*. London: Penguin.

Rayfuse, R. (2020). Out of sight, out of mind: The challenge of regulating high seas fisheries. In E. Probyn, K. Johnston, & N. Lee (Eds.), *Sustaining seas: Oceanic space and the politics of care* (pp. 141–153). London: Rowman & Littlefield.

Readfearn, G. (2020). Murray-Darling: Thousands of fish have died in NSW in the past two weeks. *The Guardian.* Retrieved from https://www.theguardian.com/australia-news/2020/feb/02/murray-darling-thousands-of-fish-have-died-in-nsw-in-past-two-weeks

Reid, T. (2020). 2020: The year of reckoning, not reconciliation: It's time to show up. *Griffith Review, 67.* Retrieved from https://www.griffithreview.com/articles/2020-year-of-reckoning/

Reiley, L. (2019a). Inside the little-known world of flavorists, who are trying to make plant-based meat taste like the real thing. *The Washington Post.* Retrieved from https://www.washingtonpost.com/business/2019/11/04/inside-little-known-world-flavorists-who-are-trying-make-plant-based-meat-taste-like-real-thing/

Reiley, L. (2019b). Veggie burgers were living an idyllic little existence. Then they got caught in a war over the future of meat. *The Washington Post.* Retrieved from https://www.washingtonpost.com/business/2019/08/25/veggie-burgers-were-living-an-idyllic-little-existence-then-they-got-caught-war-over-future-meat/

Reinhardt, M. (2015). Sprit food: A multi-dimensional overview of the decolonising diet project. In E. S. Huaman & B. Sriraman (Eds.), *Indigenous innovation* (pp. 81–105). Boston, MA: Sense Publishers.

Reisner, M. (1986). *Cadillac desert: The American west and its disappearing water.* New York, NY: Penguin.

Reverter, M., Sarter, S., Caruso, D., Avarre, J., Combe, M., Pepey, E., ... Gozlan, R. E. (2020). Aquaculture at the crossroads of global warming and antimicrobial resistance. *Nature Communications, 11*, 1870.

Rich, N. (2018). The most honest book about climate change yet. *The Atlantic*. Retrieved from https://www.theatlantic.com/magazine/archive/2018/10/william-vollmann-carbon-ideologies/568309/

Riches, G. (2018). *Food bank nations: Poverty, corporate charity, and the right to food*. Abingdon: Routledge.

Rigaud, K., de Sherbinin, A., Jones, B., Bergmann, J., Clement, V., Ober, K., ... Midgley, A. (2018). *Groundswell: Preparing for internal climate migration*. Washington, DC: The World Bank.

Ritchie, H., & Roser, M. (2020). Environmental impacts of food production. *OurWorldInData.org*. Retrieved from https://ourworldindata.org/environmental-impacts-of-food

Robin, M.-M. (2010). *The world According to Monsanto: pollution, corruption, and the control of the world's food supply*. New York, NY: The New Press.

Robinson, M. (2016). Is the moose still my brother if we don't eat him?. In J. Castricano & R. R. Simonsen (Eds.), *Critical perspectives on veganism* (pp. 261–284). New York, NY: Palgrave.

Rose, D. B. (1996). *Nourishing terrains: Australian aboriginal views of landscape and wilderness*. Canberra, ACT: Australian Heritage Commission.

Rose, D. B. (2019). Shimmer: When all you love is being trashed. In H. S. Anna Tsing, E. Gan, & N. Bubandt (Eds.),

Arts of living on a damaged planet (pp. G51–G63). Minneapolis, MN: University of Minnesota.

Rosset, P. (2013). Re-thinking agrarian reform, land and territory in *La Vía Campesina*. *Journal of Peasant Studies*, *40*(4), 721–775.

Rowen (2018). Seed rematriation. Retrieved from https://sierraseeds.org/seed-rematriation/

Rundgren, G. (2016). Food: From commodity to commons. *Journal of Agricultural and Environmental Ethics*, *29*, 103–121.

Rust, J. M. (2019). The impact of climate change on extensive and intensive livestock production systems. *Animal Frontiers*, *9*(1), 20–25.

Salazar, E., Billing, S., & Breen, M. (2020). We need to talk about chicken. Retrieved from https://www.eating-better.org/uploads/Documents/2020/EB_WeNeedToTalkAbout Chicken_Feb20_A4_Final.pdf

Salcedo Fidalgo, H. (2020). The coronavirus pandemic: A critical reflection on corporate food patterns. In *Right to food and nutrition watch. Overcoming ecological crises: Reconnecting food, nature and human rights* (pp. 16–23). Heidelberg: FIAN International & Brot für die Welt.

Samora, R. (2019). Brazil farmers push traders to end Amazon soy moratorium. *Reuters*. Retrieved from https://www.reuters.com/article/us-brazil-soybeans-moratorium/brazil-farmers-push-traders-to-end-amazon-soy-moratorium-idUSKBN1XF2J6

Samuel, G. (2018). Independent review of the Grocery Code of Conduct. Retrieved from https://treasury.gov.au/sites/default/files/2019-03/Independent-review-of-the-Food-and-Grocery-Code-of-Conduct-Final-Report.pdf

Sankey, L. (2020). Farmworkers in America's tomato capital may have one of the highest rates of coronavirus infection globally. *Over the Counter*. Retrieved from https://thecounter.org/immokalee-florida-farmworkers-covid-19-cases-coronavirus-testing-contact-tracing/amp/?__twitter_impression=true

Sassen, S. (2014). *Expulsions: Brutality and complexity in the global economy*. Cambridge, MA: Harvard University Press.

Saunders, M. (2016). Resource connectivity for beneficial insects in landscapes dominated by monoculture tree crop plantations. *International Journal of Agricultural Sustainability*, 41(1), 82–99.

Schatzker, M. (2015). *The Dorito Effect: The surprising new truth about food and flavour*. New York, NY: Simon & Schuster.

Schlosser, E. (2002). *Fast food nation: What the all-American meal is doing to the world*. London: Penguin.

Schneider, M. (2019). China's global meat industry: The world-shaking power of industrialising pigs and pork in China's Reform Era. In B. Winders & E. Ransom (Eds.), *Global meat: Social and environmental consequences of the expanding meat industry* (pp. 31–54). Cambridge, MA: The MIT Press.

Schutter, D. (2012). *Ocean grabbing as serious a threat as land grabbing*. New York, NY: UN.

Schwartz, J. D. (2013). *Cows save the planet and other improbable ways of restoring soil to heal the Earth*. White River Junction, VT: Chelsea Green Publishing.

Scrinis, G. (2013). *Nutritionism: The science and politics of dietary advice*. Chichester: Columbia University Press.

Seccombe, M. (2020, April 11–17). Focus pulling. *The Saturday Paper*, p. 8–9.

Servigne, P., & Stevens, R. (2020). *How everything can collapse*. Cambridge: Polity Press.

Seufert, P. (2020). We are nature! Human rights, envrionmental law, and the illusion of separation. In *Right to food and nutrition watch. Overcoming ecological crises: Reconnecting food, nature and human rights* (pp. 6–15). Heidelberg: FIAN International & Brot für die Welt.

Sexton, A., Garnett, T., & Lorimer, J. (2019). Framing the future of food: The contested promises of alternative proteins. *ENE: Nature and Space*, 2(1), 47–72.

Shaffer, J. M., Shaffer, R. A., Lindsay, S. P., Araneta, M. R. G., Raman, R., & Fowler, J. H. (2017). International chicken trade and increased risk for introducing or reintroducing highly pathogenic avian influenza A (H5N1) to uninfected countries. *Infectious Disease Modelling*, 2(4), 412–418.

Sharma, S., Thind, S. S., & Kaur, A. (2015). In vitro meat production: Why and how? *Journal of Food Science & Technology*, 52(12), 7599–7607. doi:10.1007/s13197-015-1972-3

Sharp, E. L. (2020). Free fish heads: A case study of knowing and practicing seafood differently. In E. Probyn, K. Johnston, & N. Lee (Eds.), *Sustaining seas: Oceanic space and the politics of care* (pp. 125–138). London: Rowman & Littlefield.

Shearman, D. (2020). COVID-19 has our attention, but other emergencies are destroying our foundations. *Croakey*. Retrieved from https://croakey.org/category/public-health-and-population-health/environmental-health/

Shiva, V. (1993). *Monocultures of the mind: Perspectives on biodiversity and biotechnology*. London: Zed Books.

Shiva, V. (2007). Biodiversity, intellecutal property rights and globalization. In B. De Sousa Santos (Ed.), *Another knowledge is possible: Beyond northern epistemologies* (Vol. 3, pp. 272–287). London: Verso.

Shiva, V. (2008). *Soil not oil: Environmental justice in an age of climate crisis*. Berkeley, CA: North Atlantic Books.

Shiva, V. (2016). *Stolen harvest: The hijacking of the global food supply*. Lexington, KY: University Press of Kentucky.

Shotwell, A. (2016). *Against purity: Living ethically in compromised times*. Minneapolis, MN: University of Minnesota Press.

Shoup, M. E. (2019). Survey: A rising generation of food consumers embrace new food technologies. *Foodnavigator-USA*. Retrieved from https://www.foodnavigator-usa.com/Article/2019/11/01/Survey-A-rising-generation-of-food-consumers-embrace-new-food-technologies?utm_source=newsletter_daily&utm_medium=email&utm_campaign=01-Nov-2019&c=16HhEPChu5Aqg9iV0Z9HLT65hbqfana4&p2=

Shove, E. (2010). Beyond the ABC: Climate change policy and theories of social change. *Environment and Planning A, 42*(6), 1273–1285.

Siegel, D. A. M. (2016). Sponsorship of national health organisations by two major soda companies. *American Journal of Preventative Medicine, 52*(1), 20–30.

Simons, M. (2020a). *Quarterly Essay 77: Cry me a river: The tragedy of the Murray-Darling Basin*. Carlton, VIC: Black Inc.

Simons, M. (2020b, May 2). Food for thought. *The Saturday Paper*, p. 8–9.

Simpson, J. (2016). Almond industry expansion. Retrieved from https://www.dpi.nsw.gov.au/__data/assets/pdf_file/0004/586435/almond-industry-expansion.pdf

Sinclair, U. (1906). *The jungle*. New York, NY: Doubleday.

Singer, P. (2006). *The ethics of what we eat: Why our food choices matter*. Emmaus, PA: Rodale.

SlowFood. (2020). The COVID-19 crisis should be the time to rethink food and farming models. Retrieved from https://www.slowfood.com/the-covid-19-crisis-should-be-the-time-to-rethink-food-and-farming-models/

Smith, G., & Kayama, R. (2020). Kenya's pastoralists face hunger and conflict as locust plague continues. *The Guardian*. Retrieved from https://www.theguardian.com/global-development/2020/may/15/kenyas-pastoralists-face-hunger-and-conflict-as-locust-plague-continues

Smith, L. T. (2012). *Decolonising methodologies: Research and Indigenous peoples*. London: Zed.

Solnit, R. (2009). *A paradise built in hell: The extraordinary communities that arise in disaster*. New York, NY: Penguin.

Solyent. (2019). Our products: We thought about your food so you wouldn't have to. Retrieved from https://soylent.com/pages/about-soylent

Souder, W. (2012). *On a farther shore: The life and legacy of Rachel Carson*. New York, NY: Broadway Books.

Southey, F. (2019a). Could yeast be a viable alternative to soy and pea protein? 'It has all the essential amino acids, is readily available and affordable'. *Nutraingredients.com*. Retrieved

from https://www.nutraingredients.com/Article/2019/10/29/
FFW-develops-yeast-protein-food-as-alternative-to-soy-and-
pea-protein?utm_source=newsletter_daily&utm_medium=
email&utm_campaign=30-Oct-2019&c=16HhEPChu
5ClEzP69jG%2BRy7k3bIPI07o&p2=

Southey, F. (2019b). Vegan start-up takes on mozzarella,
parmesan and brie: 'It's all about the classics'.
Foodnavigator.com. Retrieved from https://
www.foodnavigator.com/Article/2019/10/10/Vegan-
mozzarella-parmesan-and-brie-Food-tech-takes-on-classic-
cheese-market?utm_source=newsletter_daily&utm_
medium=email&utm_campaign=10-Oct-2019&c=
16HhEPChu5BJoBkAjo%2Fy4SstSNwS6OYo&p2=

Spadotto, B. R., Sawelijew, Y. M., Frederico, S., & Pitta, F. T.
(2020). Unpacking the finance-farmland nexus: Circles of
cooperation and intermediaries in Brazil. *Globalizations.* doi:
10.1080/14747731.2020.1766918

Springmann, M., Clark, M., Mason-D'Croz, D., Wiebe, K.,
Bodirsky, B. L., Lassaletta, L., … Willett, W. (2018). Options
for keeping the food system within environmental limits.
Nature, 562, 519–525.

State Library of Queensland. (2019). Australian South Sea
Islanders. Retrieved from https://www.slq.qld.gov.au/
discover/exhibitions/australian-south-sea-islanders

Statista. (2020). Leading countries of destination of soybean
exports from Brazil in 2019, by export value share. *Statista.*
Retrieved from https://www.statista.com/statistics/721259/
value-share-soybean-exports-brazil-country-destination/

Steel, C. (2020). *Sitopia: How food can save the world.*
London: Random House.

Steffen, W., Broadgate, W., Deutsch, L., Gaffney, O., & Ludwig, C. (2015). The trajectory of the Anthropocene: The great acceleration. *The Anthropocene Review*, 2(1), 81–98.

Stengers, I. (2015). *In catastrophic times: Resisting the coming barbarism*. Paris: Open Humanities Press.

Sun, Z., Scherer, L., Tucker, A., & Behrens, P. (2020). Linking global crop and livestock consumption to local production hotspots. *Global Food Security*, 25(100323), 1–9.

Suppan, S. (2017). Do no harm? NAFTA, non-regulation of GMOs and Mexican agriculture. *Institute for Agriculture and Trade Policy*. Retrieved from https://www.iatp.org/blog/201901/do-no-harm-nafta-non-regulation-gmos-and-mexican-agriculture

Swinburn, B. A., Kraak, V. I., Allender, S., Atkins, V. J., Baker, P. I., Bogard, J. R., … Dietz, W. H. (2019). The global syndemic of obesity, undernutrition, and climate change: The Lancet Commission Report. *The Lancet*, 393, 791–846.

Swyngedouw, E., & Ernstson, H. (2018). Interrupting the Anthropo-obScene: Immuno-biopolitics and depoliticising more-than-human ontologies in the Anthropocene. *Theory, Culture & Society*, 35(6), 3.

Tandoh, R. (2019). *Eat up! Food, appetite and eating what you want*. London: Profile Books Ltd.

Taylor, M. (2015). *The political ecology of climate change adaptation: Livelihoods, agrarian change and the conflicts of development*. Abingdon: Routledge.

Taylor, A. (2019). *Democracy may not exist, but we'll miss it when it's gone*. London: Verso.

Thill, S. (2017). Largest-ever Gulf dead zone reveals stark impacts of industrial agriculture. *Civil Eats*. Retrieved from https://civileats.com/2017/08/03/largest-ever-gulf-dead-zone-reveals-stark-impacts-of-industrial-agriculture/

Thompson, B. M. R. (2019). Aboriginal voices are missing from the Murray-Darling Basin crisis. *The Conversation.* Retrieved from https://theconversation.com/aboriginal-voices-are-missing-from-the-murray-darling-basin-crisis-110769

Thorndike, A. N., & Sunstein, C. R. N. (2017). Obesity prevention in the supermarket - choice architecture and the supplemental nutrition assistance program. *American Journal of Public Health*, 107(10), 1582–1583.

TNI. (2015a). State of Power 2015: An annual anthology on global power and resistance. Retrieved from https://www.tni.org/files/download/tni_state-of-power-2015.pdf

TNI. (2015b). The global ocean grab. Retrieved from http://worldfishers.org/wp-content/uploads/2015/01/The_Global_Ocean_Grab-EN.pdf

Todd, Z. (2018). Refracting the state through human-fish relations: Fishing, Indigenous legal orders and colonialism in north/western Canada. *Decolonization: Indigeneity, Education & Society*, 7(1), 60–75.

Tramel, S. F. (2020). Converging to overcome crisis and change the system. In *Right to food and nutrition watch. Overcoming ecological crises: Reconnecting food, nature and human rights* (pp. 24–37). Heidelberg: FIAN International & Brot für die Welt.

Trase. (2018a). New data on Trase shows soy trade from Brazil's Cerrado driving climate emissions. Retrieved from https://medium.com/trase/new-data-on-trase-shows-soy-trade-from-brazils-cerrado-driving-climate-emissions-10cc949a04c4

Trase. (2018b). Sustainability in forest-risk supply chains: Spotlight on Brazilian soy. *Transparency for Sustainable Economies*. Retrieved from https://yearbook2018.trase.earth/

Tree, I. (2018). If you want to save the world, veganism isn't the answer. *The Guardian*. Retrieved from https://www.theguardian.com/commentisfree/2018/aug/25/veganism-intensively-farmed-meat-dairy-soya-maize

Tsing, A. L. (2005). *Friction: An ethnography of global connection*. Princeton, NJ: Princeton University Press.

Tsing, A. L. (2015). *The mushroom at the end of the world: On the possibility of life in capitalist ruins*. Princeton, NJ: Princeton University Press.

Tubb, C. & Seba, T. (2019). *Rethinking food and agriculture 2020–2030*. Retrieved from https://static1.squarespace.com/static/585c3439be65942f022bbf9b/t/5d7fe0e83d119516bfc0017e1568661791363/RethinkX1Food1and1Agriculture1Report.pdf

Tuck, E., & Yang, K. W. (2012). Decolonization is not a metaphor. *Decolonization: Indigeneity, Education & Society*, *1*(1), 1–40.

Tuomisto, H. L., & Teixeira de Mattos, M. J. (2011). Environmental impacts of cultured meat production. *Environmental Science and Technology*, *45*(14), 6117–6123.

Turner, N. J. (2011). The ethnobotany of edible seaweed (Porphyra abbottae and related species; Rhodophyta: Bangiales) and its use by First Nations on the Pacific Coast of Canada. *Canadian Journal of Botany*, *81*, 283–293.

Turzi, M. (2011). The soybean republic. *Yale Journal of International Affairs*, *6*, 59–68.

Tyson. (2019a). Tyson foods unveils alternative protein products and new Raised and Rooted brand. Retrieved from https://www.tysonfoods.com/news/news-releases/2019/6/tyson-foods-unveils-alternative-protein-products-and-new-raised-rootedr

Tyson. (2019b). Tyson venture invests in New Wave Foods. Retrieved from https://www.tysonfoods.com/news/news-releases/2019/9/tyson-ventures-invests-new-wave-foods

Uluru Statement from the Heart. (2017). Retrieved from https://ulurustatement.org/

UN. (2015). Sustainable development goals. Retrieved from https://sustainabledevelopment.un.org/?menu=1300

UN. (2018). The tale of a disappearing lake. Retrieved from https://www.unenvironment.org/news-and-stories/story/tale-disappearing-lake

UN. (2019). Migration and the Climate Crisis: The UN's search for solutions. Retrieved from https://news.un.org/en/story/2019/07/1043551

UNDP. (2011). Human development report 2011: Sustainability and equity: A better future for all. Retrieved from http://hdr.undp.org/en/content/sustainability-and-equity-better-future-all

UNDP. (2019). *Beyond income, beyond averages, beyond today: Inequalities in human development in the 21st century.* New York, NY: UNDP.

UNEP. (2020a). Six nature facts related to COVID-19. Retrieved from https://www.unenvironment.org/news-and-stories/story/six-nature-facts-related-coronaviruses

UNEP. (2020b). Emerging zoonotic diseases and links to ecosystem health. Retrieved from https://www.unenvironment.

org/resources/emerging-zoonotic-diseases-and-links-ecosystem-health-unep-frontiers-2016-chapter

UNFCCC. (2019). Impossible foods: Creating plant-based alternatives to meat | Singapore, Hong Kong, USA, Macau. Retrieved from https://unfccc.int/climate-action/momentum-for-change/planetary-health/impossible-foods

Unión de Trabajadores Agricolas Fronterizos. (2020). Coronavirus hits the migrant farm workers' border farm workers project. Retrieved from https://viacampesina.org/en/covid-19-several-members-of-la-via-campesina-highlight-the-vulnerability-of-peasants-and-workers/

van der Ploeg, J. D. (2008). *The new peasantries: Struggles for autonomy and sustainability in an era of empire and globalisation*. Oxford: Earthscan.

van der Ploeg, J. D. (2020). From biomedical to politicoeconomic crisis: The food system in times of Covid-19. *The Journal of Peasant Studies, 47*(5), 944–972. doi:10.1080/03066150.2020.1794843

van der Weele, C., Feindt, P., Goot, A. J., Van Mierlo, B., & Boekel, M. (2019). Meat alternatives: An integrative comparison. *Trends in Food Science and Technology, 88*, 505–512. doi:10.1016/j.tifs.2019.04.018

Venturini, F., Değirmenci, E., & Morales, I. (Eds.). (2019). *Social ecology and the right to the city*. Montréal: Black Rose Books.

Venturini, F. (2019). Reconceptualising the right to the city and spatial justice through social ecology. In F. Venturini, E. Değirmenci, & I. Morales (Eds.), *Social ecology and the right to the city* (pp. 86–100). Montréal: Black Rose Books.

Vergés, A., Crawford, M., Kajlick, L., Marzinelli, E. M., Söderlund, A., Steinberg, P. D., ... Campbell, A. H. (2020). Operation Crayweed: Merging art and science to restore underwater forests. In E. Probyn, K. Johnston, & N. Lee (Eds.), *Sustaining seas: Oceanic space and the politics of care* (pp. 237–252). London: Rowman & Littlefield.

Vernon, C. (2019). In conversation with Bruce Pascoe. Retrieved from https://greatersydneylandcare.org/in-conversation-with-bruce-pascoe/

Vidal, J. (2013). Millions face starvation as world warms, say scientists. *The Guardian*. Retrieved from https://www.theguardian.com/global-development/2013/apr/13/climate-change-millions-starvation-scientists

Vince, G. (2020). Root and branch. *The Guardian Weekly*, May 22, 202(23), 8–11.

Vittuari, M., De Menna, F., & Pagani, M. (2016). The hidden burden of food waste: The double energy waste in Italy. *Energies*, 9, 660. doi:10.3390/en9080660

Vivero-Pol, J., Ferrando, T., De Schutter, O., & Mattei, U. (Eds.). (2019). *Routledge handbook of food as a commons*. London: Routledge.

Vivero-Pol, J. (2017). Food as commons or commodity? Exploring the links between normative valuations and agency in food transition. *Sustainability*, 9, 422.

Vollmann, W. T. (2018). *No immediate danger: Volume One of Carbon Ideologies*. New York, NY: Viking.

Wahlquist, C. (2020). Australia's summer bushfire smoke killed 445 and put thousands in hospital, inquiry hears. *The Guardian*. Retrieved from https://www.theguardian.com/australia-news/2020/may/26/australias-summer-bushfire-smoke-killed-445-and-put-thousands-in-hospital-inquiry-hears#maincontent

Wallace, R. (2016). *Big farms make big flu: Dispatches on infectious disease.* New York, NY: Monthly Review Press.

Wallace-Wells, D. (2019). *The uninhabitable Earth: A story of the future.* London: Penguin.

Walsh, C. (2010). Development as *Buen Vivir*: Institutional arrangements and (de) colonial entanglements. *Development,* 53(1), 15–21.

Walter, P. (2012). Educational alternatives in food production, knowledge and consumption: The public pedagogies of *Growing Power and Tsyunhehkw^*. *Australian Journal of Aduclt Learning, 52*(3), 574–594.

Waterhouse, J., Schaffelke, B., Bartley, R., Eberhard, R., Brodie, J., Star, M., & Kroon, F. (2017). *2017 scientific consensus statement: Land use impacts on Great Barrier Reef water quality and ecosystem condition.* Retrieved from https://www.reefplan.qld.gov.au/__data/assets/pdf_file/0029/45992/2017-scientific-consensus-statement-summary.pdf

Watson, E. (2019a). 'Cultivated' meat could be the most consumer-friendly term for cell-cultured meat, suggests Mattson/GFI research. *Food Navigator-USA.com.*

Watson, E. (2019b). Selling plant-based and cell-cultured meat will be about 'benefits for me' says Hartman Group. *Food Navigator - USA.com.* Retrieved from https://www.foodnavigator-usa.com/Article/2019/07/24/Selling-plant-based-and-cell-cultured-meat-will-be-about-benefits-for-me-says-Hartman-Group?utm_source=EditorsSpotlight&utm_medium=email&utm_campaign=2019-09-30&c=16HhEPChu5DxL4ZU1Szl3peQYFlCnqKS

Watts, V. (2013). Indigenous place-thought and agency amongst humans and non-humans (First Woman and Sky Woman go on a European tour!). *Decolonization, Indigeneity, Education and Society*, 2(1), 20–34.

Watts, J. (2020). Oil and gas firms 'have had far worse impact than first thought'. *The Guardian*. Retrieved from https://www.theguardian.com/environment/2020/feb/19/oil-gas-industry-far-worse-climate-impact-than-thought-fossil-fuels-methane?CMP=share_btn_link

Waziyatawin, A. W., & Yellow Bird, M. (Eds.). (2005). *For Indigenous eyes only: A decolonization handbook*. Santa Fe, NM: School of American Research Press.

Weis, T. (2007). *The global food economy: The battle for the future of farming*. London: Zed Books.

Weis, T. (2013). *The ecological hoofprint: The global burden of industrial livestock*. London: Zed Books.

Wells, S. (2013). Greenwashing or real progress for animals?. *Huffington Post*. Retrieved from https://www.huffpost.com/entry/greenwashing-or-real-prog_b_3503137

WFP. (2020). COVID-19: Potential impact on the world's poorest people. Retrieved from https://docs.wfp.org/api/documents/WFP-0000114040/download/

WHO. (1986). Ottawa charter for health promotion. Retrieved from https://www.who.int/healthpromotion/conferences/previous/ottawa/en/

WHO. (2011). Rio political declaration on social determinants of health. Retrieved from https://www.who.int/sdhconference/declaration/Rio_political_declaration.pdf

WHO. (2014). Quantitative risk assessment of the effects of climate change on selected causes of death, 2030s and 2050s.

Retrieved from https://www.who.int/globalchange/
publications/quantitative-risk-assessment/en/

WHO. (2019). Toward healthier populations: A new vision -
Technical Briefing at WHA72. Geneva: WHO. Retrieved from
https://www.youtube.com/watch?v=bpjN40Eh3PE

WHO. (2020). About social determinants of health. Retrieved
from https://www.who.int/social_determinants/
sdh_definition/en/

Wiedmann, T., Lenzen, M., Keyßer, L. T., & Steinberger, J. K.
(2020). Scientists' warning on affluence. *Nature
Communications*, *11*, 3107.

Wilkinson, R., & Pickett, K. (2009). *The spirit level: Why
equality is better for everyone*. London: Penguin.

Wilkinson, R., Pickett, K. E., & De Vogli, R. (2010). Equality,
sustainability, and quality of life. *BMJ Global Health*, *341*,
1138–1140.

Willet, W., Rockström, J., Loken, B., Springmann, M., Lang,
T., Vermeulen, S., … Murray, C. J. L. (2019). Food in the
Anthropocene: The EAT-Lancet Commission on healthy diets
from sustainable food systems. *The Lancet*, *393*(10170),
447–492.

Williams, T., & Hardison, P. (2013). Culture, law, risk and
governance: Contexts of traditional knowledge in climate
change adaptation. In J. K. Maldonado, B. Colombi, & R.
Pandya (Eds.), *Climate change and Indigenous peoples in the
United States* (pp. 23–36). Cham: Springer.

Winders, W., & Ransom, E. (Eds.). (2019). *Global meat:
Social and environmental consequences of the expanding meat
industry*. Cambridge, MA: The MIT Press.

Wise, T. (2020). Failing Africa's farmers: New report show Africa's Green Revolution is "failing on its own terms". *Institute for Agriculture and Trade Policy*. Retrieved from https://www.iatp.org/blog/202007/failing-africas-farmers-new-report-shows-africas-green-revolution-failing-its-own-terms

Wolf Ditkoff, S., & Grindle, A. (2017). Audacious philanthropy. *Harvard Business Review*. Retrieved from https://hbr.org/2017/09/audacious-philanthropy

Wolford, W., Borras, S. M., Jr, Hall, R., Scoones, I., & White, B. (2013). Governing global land deals: The role of the state in the rush for land. *Development and Change*, 44(2), 189–210.

World Forum of Fisher Peoples. (2015). No to blue carbon, yes to food sovereignty and climate justice!. Retrieved from http://worldfishers.org/wp-content/uploads/2015/12/Blue_Carbon_Statement.pdf

WorldFish. (2017). From local to global: How research enables resilient and sustainable small-scale fisheries. Retrieved from https://www.worldfishcenter.org/pages/from-local-to-global-how-research-enables-resilient-sustainable-small-scale-fisheries/

Wrenn, C. L. (2019). The Vegan Society and social movement professionalization, 1944-2017. *Food and Foodways*, 27(3), 190–210.

Wright, C., & Nyberg, D. (2015). *Climate change, capitalism, and corporations: Processes of creative self-destruction*. Cambridge: Cambridge University Press.

Yaffe-Bellany, D. (2019). The new makers of plant-based meat? Big meat companies. Retrieved from https://www.nytimes.com/2019/10/14/business/the-new-makers-of-plant-based-meat-big-meat-companies.html

Yazzie, M. K., & Balday, C. R. (2018). Introduction: Indigenous peoples and the politics of water. *Decolonisation: Indigeneity, Education & Society*. 7(1), 1–18.

Young, N. (2018). Overfished West Coast hoki fishery collapses. *Greenpeace*. Retrieved from https://www.greenpeace.org/new-zealand/press-release/overfished-west-coast-hoki-fishery-collapses/

Yunkaporta, T. (2019). *Sand talk: How Indigenous thinking can save the world*. Melbourne: Text Publishing.

Zhou, C., & Walsh, M. (2020). Australia pledged to 'step up' in the Pacific amid growing Chinese influence, but are we on track? *ABC News*. Retrieved from https://www.abc.net.au/news/2020-01-18/australia-pacific-step-up-in-review/11863150

INDEX